TPM Implementation

Other McGraw-Hill Books of Interest

IRESON, COOMBS • *Handbook of Reliability Engineering and Management*

KAEWERT, FROST • *Developing Expert Systems for Manufacturing*

LUBBEN • *Just-in-Time Manufacturing*

MAYNARD • *Industrial Engineering Handbook*

NEVINS, WHITNEY • *Concurrent Design of Products and Processes*

STARK • *Managing CAD/CAM*

TEICHOLZ, ORR • *Computer Integrated Manufacturing Handbook*

CLELAND, BIDANDA • *The Automated Factory Handbook: Technology and Management*

TOMPKINS • *Winning Manufacturing*

FEIGENBAUM • *Total Quality Control*

JURAN • *Juran's Quality Control Handbook*

SLATER • *Integrated Process Management: A Quality Model*

TAYLOR • *Optimization and Variation Reduction in Quality*

ROSS • *Taguchi Techniques for Quality Engineering*

RAHEJA • *Assurance Technologies*

OTT, SCHILLING • *Process Quality Control*

TPM Implementation

A Japanese Approach

Masaji Tajiri

Fumio Gotoh

McGraw Hill Education (India) Private Limited
NEW DELHI

McGraw Hill Education Offices
New Delhi New York St Louis San Francisco Auckland Bogotá Caracas
Kuala Lumpur Lisbon London Madrid Mexico City Milan Montreal
San Juan Santiago Singapore Sydney Tokyo Toronto

McGraw Hill Education (India) Private Limited

McGraw Hill Edition (India) Edition 2014

RDDZLCRYUDZAA

Reprinted in India by arrangement with The McGraw-Hill Companies, Inc., New York

Sales territories: India, Pakistan, Nepal, Bangladesh, Sri Lanka and Bhutan

Library of Congress Cataloging-in-Publication Data

Goto, Fumio
(TPM suishin no pointo, jishu hozen nanatsu no suteppu. English)
TPM implementation, a Japanese approach / Masaji Tajiri, Fumio Gotoh.
p. cm.
Translation of: TPM suishin no pointo, jishu hozen nanatsu no suteppu / Gato Fumio, Tajiri Masaji.
ISBN 0-07-062834-3
1. Plant maintenance—Japan. I. Tajiri Masaji, Date.
II. Title.
TS192.G68513 1992
658.2'02—dc20 92-9038
CIP

ISBN-13: 978-93-392-1830-0
ISBN-10: 93-392-1830-2

Published by McGraw Hill Education (India) Private Limited,
P-24, Green Park Extension New Delhi 110 016, and printed at
Sanat Printers, 312, EPIP, Kundli, Dist. Sonipat, Haryana 131 028

To those unknown frontline personnel who have contributed to the creation of TPM concepts on the manufacturing shopfloor

Contents

Preface

In order to maximize the effectiveness of equipment throughout its entire life, Total Productive Maintenance (TPM), which is based primarily on the productive maintenance (PM) concept imported from the United States, was initiated in 1971 and promoted by the Japan Institute of Plant Maintenance. Thereafter, TPM continued in a relatively small number of factories without achieving remarkable success. In those days, Japanese manufacturers were faced with severe economic challenges caused by a series of oil crises. Therefore, they were searching seriously for effective measures to survive in the marketplace.

In the meantime, a step-by-step small group activity was begun in the Chuo Spring Company in the late 1970s. In 1981 a prototype of the seven-step program to implement an operator's routine maintenance system on the shopfloor was developed in the Tokai Rubber Industries and produced significant benefits. Since then, the number of factories implementing the TPM system has been growing year-by-year, because the system actually achieves improved operating conditions in existing plants and increases employees' knowledge and skill. TPM concepts have been expanding continuously with the accumulation of wisdom created on the shopfloor.

Japanese manufacturers, furthermore, faced other serious difficulties in business, such as sudden and frequent changes in oil prices and currency exchange rates. Many of them found an escape from the challenging problems of those trying times through the automation of repetitive manual work that was made possible by significant progress in microelectronic and computer technologies. Therefore, they have succeeded in maintaining their position in the worldwide marketplace. Under these circumstances, TPM, since the late 1980s, has been rapidly recognized by a growing number of companies as a highly effective methodology for dealing with matters of not only plant maintenance, but also plant engineering and product design. The TPM of today may be viewed more aptly as an abbreviation for Total Production Management, rather than for Total Productive Maintenance.

This book discusses the results of the persistent efforts made by frontline personnel, mainly on the equipment-oriented and manual work-oriented shopfloor. It also includes the first presentation of a concrete approach for attaining Zero Defects through process quality assurance. This book is, therefore, appropriately dedicated to all the frontline managers, engineers, operators, maintenance personnel, and other employees who have given their greatest efforts to implement the TPM system on the shopfloor; and who, even today, continue to enhance it.

The authors also would like to express their deepest gratitude to the Japanese companies listed here which willingly gave their permission to publish the technical experience and know-how obtained through TPM activities in their organizations even though specific references to them do not appear in this book.

- Asahi Glass Company, Ltd., Sagami Works
- Bridgestone Corporation, Hikone Plant
- Chuo Spring Company, Ltd., Hekinan Plant
- Daihatsu Motor Company, Ltd.
- Hitachi, Ltd., Takasaki Plant
- SOMIC Ishikawa, Inc.
- Tokai Rubber Industries, Ltd., Komaki Plant
- Tokico, Ltd., Sagami Plant
- Toyoda Automatic Loom Works, Ltd.
- Toyota Auto Body Company, Ltd.

The authors hope the good will and experience of these companies can facilitate the discovery of an expeditious solution to the challenges which currently face manufacturers in the United States.

Finally, the authors wish to acknowledge the significant role played by our associate, Wallace Humphreys, who transformed Masaji Tajiri's rough draft of the text into a readable presentation of the latest information from Japan about TPM.

Fumio Gotoh
Masaji Tajiri

Introduction

The Second Industrial Revolution and Automation

The first industrial revolution liberated human beings and domestic animals from burdensome physical labor. In this current decade, industrialized, capitalistic countries around the world are facing a second industrial revolution, one goal of which is to free human beings from monotonous and repetitive manual work in various work places.

This second industrial revolution, initiated during the early 1970s in the area of consumer goods, was unexpectedly triggered by severe competition, primarily among Japanese manufacturers producing small pocket calculators. The subsequent explosion of the demand for semiconductors resulted in the rapid progress of relevant technologies and the spread of microelectronic consumer goods, such as personal computers and television game machines. As a result, this trend of the times is jeopardizing the gigantic manufacturers of mainframe and minicomputers. This situation revealed itself in the 1987 New York stock market crash, Black Monday. The revolution is continuing in various aspects of human life in response to further technical progress and innovations.

This same trend makes it possible for manufacturers to apply microelectronic technologies, such as sophisticated sensors, microcomputers, numerical controls, programmable controllers, robots, etc., within feasible budgetary limits. This application, more importantly, is attainable by means of relatively easy technical skills. In keeping with this development, Japanese manufacturers, including large companies and more serious small organizations such as family-operated parts suppliers, are applying automation to repetitive manual work in order to overcome difficulties in the business environment. Automation not only frees factory workers from monotonous physical work, but also significantly improves the quality and reduces the sales price of goods. Such a trend toward automation of repetitive manual work is referred to in this book as the second industrial revolution.

American and other early industrialized Western societies, on the other hand, traditionally tend to fear and avoid automation. Even Norbert Wiener, the founder of cybernetics who accurately recognized the trend toward computerization and automation, and labeled it "the second industrial revolution," predicted, on the basis of this negative attitude, that automation will result in a large amount of unemployment.

American companies were responding to this trend by exerting major efforts to apply computerization in managerial, financial, and other administrative areas. These same efforts, however, had been resisted for a long time on the manufacturing shopfloor. It is ironic that in Japan and the other Asian NIES countries that have embraced automation, there are chronic labor shortages.

Although there are essential questions to be answered about today's mass-producing and mass-consuming society, the manufacturing business was and must continue to be a very important component in the undergirding of society, as evidenced by the history following the first industrial revolution. Production technology, including automation, must continue to be a most crucial issue, not only for manufacturers, but also for countries to become and remain competitive in the worldwide marketplace. The above viewpoint is especially relevant in light of the recent developments in the Commonwealth of Independent States (CIS) and Eastern Europe.

The Zero-Oriented Concept

During the second industrial revolution, a large number of major and minor concepts, involving quality and production technology developed in Japan, were described in various books. This phenomenon in the United States began with a reconsideration of Deming's quality concepts, which had been neglected for a long time in his own country. More specifically, what followed was a flood of information, such as Japanese-style management, TQC, the Toyota production system, and, the newcomer, TPM. Because most of the written materials are based on relatively superficial or partial understanding of these concepts by authors, little or no impact on American manufacturers is anticipated. As a result, a huge amount of abstract and empty discussions transpire whenever one of these concepts is introduced. In some extreme cases, books are put together with scissors and paste by personnel who lack expertise in these matters, and, therefore, produce only confusion among company leaders and engineers.

A term, *just in time* (JIT), for example, is a common American expression and is used even in television ads. Almost no manufacturers, nevertheless, have successfully implemented in their factory a

JIT system in the true sense, despite the fact that JIT is a simple idea prescribing that necessary quantities of necessary materials exist whenever needed. The frontline personnel of Toyota Motor, led by Taiichi Ohno, spent some decades establishing this simple system after Ohno gleaned the idea from operations and material handling in supermarkets which he visited in the United States during the early 1950s.

Another term, kanban ("signboard" in English), originated in a Toyota factory's dialect and is frequently referred to in this book. It means a tag or plate which bears the minimal information needed in terms of workpieces conveyed between processes and purchased raw materials, such as the material code and specification number, manufacturing process or vendor's name, quantity, location to be supplied, etc. A kanban is hung on or attached to necessary materials and received at a downstream process as a delivery slip or packing list, and, on return, is submitted to the source of supply as a purchase order to request additional, necessary materials. This invaluable system was created on the manufacturing shopfloor to help establish a JIT system in those days when no computer was available to facilitate material handling and production control. Even today, many factories in Japan do not use computers for these purposes.

The single reason why the implementation of a JIT system is so difficult relates to the significant difficulty in achieving Zero Defects. Any minor quality defects in supplied materials can stop an entire production line because each process has no alternatives in stock. All companies, therefore, which try to implement a JIT system based only on a superficial understanding of and in the absence of Zero Defects, certainly face factory-wide chaos on the manufacturing shopfloor.

In order to achieve Zero Defects, Zero Breakdowns must be realized as an absolute prerequisite. Zero Accidents, in turn, are not attainable without Zero Breakdowns and Zero Defects. The JIT plus TQC approach, for example, makes no sense within the above considerations because any presently existing QC concept allows for an incidence of quality defects within a certain reasonable tolerance and does not strive for Zero Defects by way of a concrete approach. Herein lies the reason why Japanese automobile manufacturers and their parts suppliers are developing TPM activities. In this regard, the Zero-oriented concept refers to a definite type of approach to searching for the absolute value, Zero, not by way of abstract armchair ideas, but by way of concrete and actual technical approaches that depend upon the total cooperation of all relevant personnel.

All concepts and strategies, based on the Zero-oriented concept described in this book, can be applied to almost all shopfloors in various types of industries. This book, therefore, draws upon typical and widely applicable materials to facilitate the understanding of TPM by a diverse audience. It, however, behooves any company or factory which is interested in trying to implement a TPM system to adapt the information presented herein to its own operating conditions and to the characteristics of its own products. For example, JIT is a very effective system under certain operating and social conditions along with the technical progress in automation. This is especially the case in a factory that manufactures products such as automobiles which have wide variations in their specifications and relatively low value added per their cubic measure.

In adapting TPM to a specific situation, frontline leaders must do so firsthand on their own equipment by following definitive instructions with special reference to Steps 1 through 3, described in this book. These same personnel thereby may come to understand thoroughly by practice the Zero-oriented concept. Abstract and empty discussions about TPM, on the contrary, are unproductive strategies.

Quality Is Not a Matter to Be Controlled

In the process of establishing a TPM system in a factory, frontline personnel are trained and encouraged to apply thorough statistical techniques, which are also utilized in conventional QC concepts. These personnel thereby are able to analyze their problems and respond with adequate strategies for achieving Zero Breakdowns and Zero Defects. The Zero-oriented concept, notwithstanding this approach, comes into conflict with the traditional avenues toward matters of quality.

Quality of products, by its very nature and from TPM's standpoint, demands that it be created by highly skilled and motivated workers, and with equipment operated and maintained under optimal conditions. Quality, from this viewpoint, is not something that can be controlled.

During this second industrial revolution, the importance of this new approach is becoming increasingly relevant as automation is brought to bear upon repetitive manual work. In other words, it is apparent that only the proper maintenance of equipment conditions produces quality which meets given specifications. Excessive efforts on predetermined quality do not represent a single adequate approach. The new approach, on the contrary, always addresses the source of undesirable conditions identified on the shopfloor. All companies implementing a TPM system, therefore, develop their own activities, under the name of TPM, simultaneously in the plant engineering and product design departments. These activities strive for the prevention of potential problems anticipated in the forthcoming commercial production of new products in the newly installed or revamped production line.

Additionally, in most Japanese factories, quality is inspected for and managed, in large part, by shopfloor workers. The quality assurance or inspection department generally takes charge of only special inspections which cannot be dealt with by the workers. Based on this same new approach, the number of inspectors also is reduced as far as possible by the training of skillful workers, the automation of manual inspection work, and the implementation of highly reliable equipment. These foregoing considerations must be well understood and appreciated before anyone can effectively approach the subject of TPM concepts.

Equipment and Human Perspectives

TPM aims at improving existing plant conditions and at increasing the knowledge and skills of frontline personnel in order to achieve Zero Accidents, Zero Defects, and Zero Breakdowns. In all factories of any country, all problems recurring on the manufacturing shopfloor result from the insufficient technical expertise of and the inadequate routine supervision by frontline managers and engineers. These problems in productivity and quality, however, are frequently ascribed to shopfloor workers. This situation, on the contrary, ultimately originates from a lack of awareness on the part of top management and higher level personnel, who do not know or even attempt to learn about the nature and characteristics of the manufacturing system and its processes.

This book addresses this situation by describing how to establish an operators' routine maintenance system by detailing a step-by-step educational program which is divided into seven steps. This program assures that operators can and will remedy existing plant conditions. Each of the prescribed steps, therefore, has aims based on both equipment and human perspectives. It is, however, also a role of frontline managers and engineers to improve and maintain productivity and quality. To do so, they must change their traditional way of thinking, which entails the implementation of a TPM system involving the thorough application of the detailed procedures described in this book. The seven-step program should, therefore, be considered an educational and training program for factory leadership which prepares it to execute the actual procedures and demonstrate the benefits of a TPM system to shopfloor personnel.

Based on the standpoint discussed in this Introduction, Chaps. 1 and 2 cover the necessity of and minimum requirement for understanding the overall TPM system. Procedures and case studies of the seven-step program, for the purpose of successful implementation of a TPM system on the manufacturing shopfloor, are detailed in Chaps. 5 through 11 and Chap. 12 in reference to equipment-oriented and manual work-oriented shopfloors, respectively. These concrete and practical presentations on two types of activities supplementing one another are preceded by Chaps. 3 and 4, which are to help company leaders and frontline personnel understand the common concepts and procedures applied in any given program.

Masaji Tajiri

Chapter

1

TPM Aims at the Elimination of Losses

1.1 The Shopfloor Suffers from Diverse Mistakes

The production department uses equipment daily to manufacture various products from raw materials. The maintenance department, on the other hand, is responsible for upkeeping equipment, even though the production department has continual, hands-on knowledge of the equipment's condition. As a result, the shopfloor is hindered by significant mistakes made by other departments, which have no appreciation of the real factors that impinge on production.

Sales department. Errors in the code number, quantity, specifications, delivery date, or destination of goods result in severe troubles for the shopfloor. If these mistakes are detected in the course of manufacturing, they may be corrected. It is too late, however, if they are discovered at the time of the customer's acceptance inspection.

Production control department. Errors in communication with the sales department result in wrong forecasts, production plans, and work orders.

Purchasing department. Errors in purchase orders result in a shortage of raw materials or the acquisition of useless materials with off-specifications.

Shipping department. Errors in the quantity of items shipped affects the customer's production line schedule.

Product design department. Product design focused on only appearance or function often neglects the need for manufacturing ease. Workers are always blamed for problems with assembly mistakes and skilled workers are involved in a lot of remedial work.

Plant engineering department. Poor plant engineering causes many troubles in commissioning and results in start-up delays. Even after initiating commercial production, workers face chronic equipment breakdowns and quality defects, as well as maintenance work difficulties.

Maintenance department. A piece of equipment when overhauled, for example, was reassembled with incorrect spare parts. Immediately after it was restarted, sporadic breakdowns occurred. The needed repair took three days.

To some extent, these mistakes are repeated in all factories. Sometimes they result from a misunderstanding of the managers' role in dealing with the problems caused by these kinds of mistakes. Most of these problems can be prevented with relatively little effort, if a suitable managerial system exists.

1.2 The Shopfloor Is Flooded with Losses

Losses, a huge quantity of which occur on the manufacturing shopfloor, are caused by other departments, as well as by operators and maintenance personnel. In TPM, time losses include equipment breakdowns, changeovers, minor stoppages, and low-speed operations, and material losses include quality defects and reduced yields.

Furthermore, these losses are classified into two cause-based categories: function-loss breakdowns and function-reduction breakdowns. Function-loss breakdowns refer to the operating conditions under which breakdowns or malfunctions stop or impede the function of equipment and result in the cessation of production. Function-reduction breakdowns refer to the operating conditions under which low-speed operations or quality defects occur, although production can continue. In these situations, the resulting off-specification products are reworked or scrapped. Among these losses, quality-related losses, which have two types of characteristics in terms of time and material, are the most significant.

From the viewpoint of recognition, losses are categorized as one of two types. The first type is one which is easily recognized by anyone. For example, in the event that a large quantity of rejects is manufactured and needs to be scrapped, the situation is a loss. This type of loss is referred to as an exposed loss.

On the other hand, the second type of losses is often neglected due to the poor technical expertise of the personnel concerned. Also, at times, they are not thought of as losses at all. For example, a workpiece clogs a chute and causes equipment to shut down or idle; or sensors detect a machine malfunction or actuate improperly, then shut equipment down automatically. This kind of interruption is referred to as a minor stoppage. When operators observe these events, they may restart normal equipment operation by simply poking at the clogged workpiece or by resetting the console. As a result, such minor losses are always overlooked. Or, equipment is operated at lower than standard operating speed, although effectiveness can be increased if speed is accelerated. Such operating conditions looked at superficially, however, seem to be maintained without trouble, and reduced equipment effectiveness goes unrecognized. These losses are referred to as hidden losses.

In other cases, material losses, such as workpiece breakages, flash, runner, and gates, which are typically found in glass and plastic moldings, and metal casting plants, are always underestimated because such raw materials can easily be recycled. Most of these losses are, therefore, overlooked and remain hidden.

Anyone, including managers, who is involved in a manufacturing business, only rarely has the opportunity to visit other companies' plants. Even then, it may be very difficult to identify their own technical levels and comprehend the quantity of their losses by making comparisons with others, due to a lack of skill in making analogies.

Generally, all personnel working under various conditions in any given plant are continually surrounded by losses. Having become accustomed from early on to accepting such losses as routine, they ignore them as losses. This kind of attitude is commonplace.

1.3 Understanding the Nature of Losses

1.3.1 Six big losses

In TPM, the relationship between losses and equipment effectiveness is clearly defined in terms of both the quality of the product and equipment availability. On the basis of a thorough examination of the factors that reduce equipment effectiveness, major losses are categorized into the following six types, as illustrated in Fig. 1.1.

1. *Breakdown losses* are caused by equipment defects which require any kind of repair. These losses, for example, consist of downtime along with the labor and spare parts required to fix the equipment; their magnitude is measured by downtime.
2. *Setup and adjustment losses* are caused by changes in operating conditions, such as the commencement of production runs or start-up at

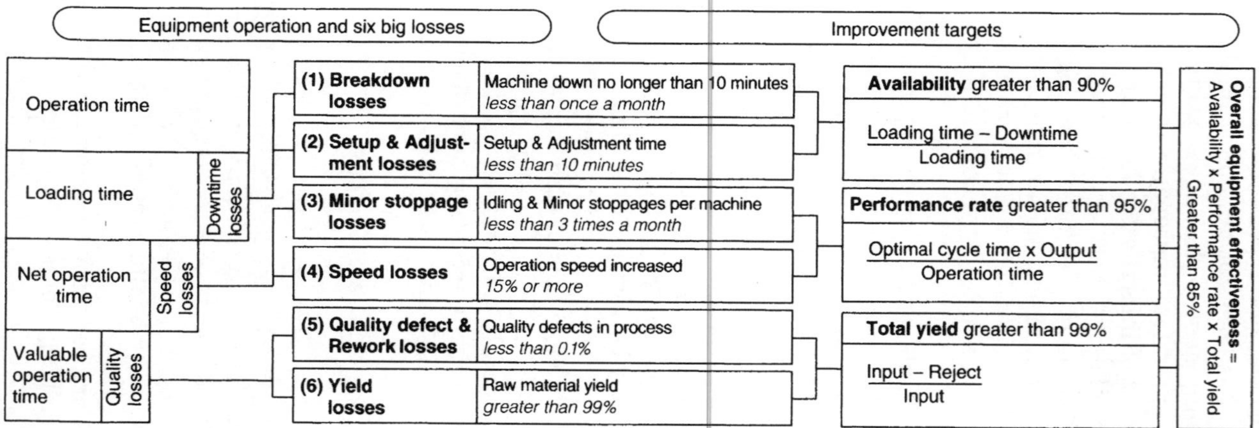

Figure 1.1 Six big losses and their improvement targets.[1]

each shift, changes in products, and conditions of operation. These losses, for example, consist of downtime, setup (equipment changeovers, exchanges of dies, jigs, and tools), start-up, and adjustment; their magnitude is also measured by downtime.

3. *Minor stoppage losses* are caused by events such as machine halting, jamming, and idling. In general, these losses cannot be recorded automatically without suitable instruments. They are, therefore, assessed based on the formula (100 percent minus performance rate). When operators cannot correct minor stoppages within a certain designated time (i.e., 10 minutes) many companies regard such minor stoppages as breakdowns in order to emphasize their importance, even though no damage has occurred to the equipment.
4. *Speed losses* are caused by reduced operating speed. Equipment cannot be operated at original or theoretical speed. At higher operating speeds, quality defects and minor stoppages frequently occur. Equipment, thereby, is required to operate at a lower moderate speed. Speed losses are measured in terms of the ratio of theoretical to actual operating speed.
5. *Quality defect and rework losses* are caused by off-specification or defective products manufactured during normal operation. These products must be reworked or scrapped. The losses consist of the labor required to rework the products and the cost of the material to be scrapped; their magnitude is measured by the ratio of quality products to total production. Sometimes, they are designated as "quality defects in process" in order to distinguish them from other quality defects like unsalable or defective products manufactured during start-up and adjustment operations.
6. *Yield losses* are caused by unused or wasted raw materials and are exemplified by the quantity of rejects, scraps, chips, etc. The yield losses are divided into two groups. One is the raw material losses resulting from product designs, manufacturing methods, and equipment restrictions, such as the flash, gate, and runner in plastic molding. The other group is the adjustment losses resulting from quality defects associated with stabilizing operating conditions at the commencement of work, changeover, etc. The longer the time spent in changeover, the less favorable the evaluation, which is determined by adding the setup and adjustment losses plus the yield losses, in terms of both time and material losses.

These six big losses are presented in three indices, i.e., availability, performance rate, and total yield rate. Their multiplicative effect is shown in Fig. 1.2 and is referred to as overall equipment effectiveness.

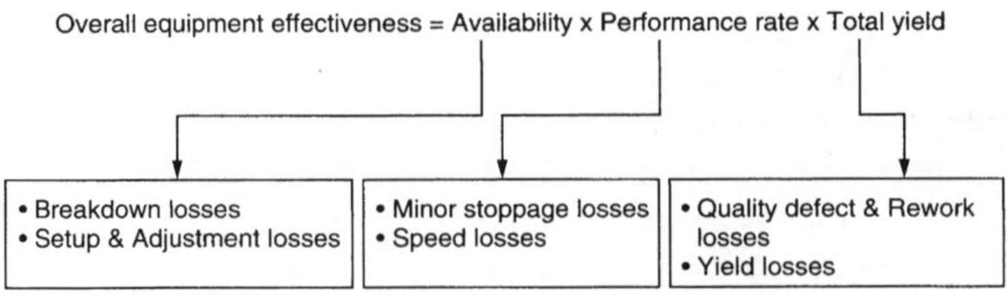

Figure 1.2 Overall equipment effectiveness.

With conventional production management concepts, it used to be impossible to ascertain totally the effectiveness of equipment, a process, or an entire plant. By introducing the concept of overall equipment effectiveness, productivity at each level is clearly defined with accurate and meaningful figures. This concept, for the first time, makes it possible to compare the trend of productivity from past to present, and from production line to production line in the same plant or plant-by-plant. Comparisons also may be made among various products manufactured by different companies. Additionally, the hidden losses, which never have been recognized as losses, are plainly and surprisingly exposed.

Besides the six big losses, many companies identify other kinds of losses, as indicated by the characteristics of their equipment and products. Elimination of all these losses is assigned the highest priority and is pursued vigorously by all of the company's constituents.

1.3.2 Chronic losses and sporadic losses

Prior to the discussion about the elimination of losses, their characteristics must be understood. In general, losses, such as breakdowns and quality defects, are discussed in terms of their occurrence. TPM, however, deals with losses based on equipment effectiveness.

The gap between the equipment's actual effectiveness and its optimal value is referred to as chronic loss when the same loss recurs in a narrow range of incidence. Sometimes, however, the recurrence of losses increases suddenly beyond the usual range, and equipment effectiveness thereby drops rapidly, as illustrated in Fig. 1.3. This kind of loss is referred to as a sporadic loss, and results from various changes in raw materials, operating conditions, jigs, tools, electric current, voltage, atmospheric temperature, humidity, air flow, misoperation, and so on.

When sporadic losses reduce equipment effectiveness, the causes are generally traced by means of troubleshooting. Then, suitable corrective actions to return it to normal conditions may be taken. On the other hand, if equipment effectiveness remains at a lower level because of chronic losses, it can never be improved by traditional countermeasures. All conventional theories and fixed ideas must be abandoned.

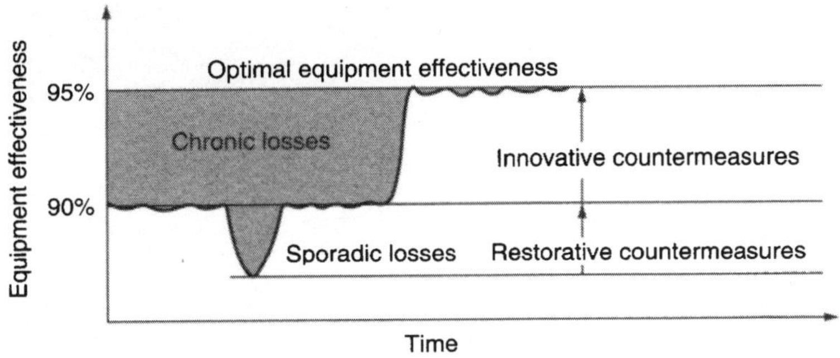

Figure 1.3 Chronic and sporadic losses.

Only innovative remedies based on breakthrough ideas can challenge the goals: Zero Breakdowns and Zero Defects.

1.4 Why Losses Occur

1.4.1 Causes of losses

The causes of losses, aside from their being sporadic or chronic, can be categorized into three types: single, multiple, and complex causes, as illustrated in Fig. 1.4.

When only one cause results in losses, it is referred to as a "single cause." A single cause, such as misoperation, improper supply of raw material, V-belt breakage, and bearing seizure, often results in sporadic losses. However, this kind of cause and its countermeasures are relatively easy to discover.

Meanwhile, several single causes may exist simultaneously, each one provoking losses triggered by minute changes in atmospheric temperature, humidity, or any other operating conditions. This kind of problem seems to be most successfully solved by taking adequate countermeasures focused on each individual cause. In time, however, another similar problem may occur at the same or a nearby location on the same piece of equipment. When causes exist simultaneously and each of them is independent of the others in its effect, they are referred to as "multiple causes."

A different scenario is one in which plural causes exist, but none of them results in losses by itself. Losses occur only when a particular combination of these causes comes into existence simultaneously and by chance, which is very seldom. This kind of situation is referred to as one of "complex causes."

All equipment is designed to work in balance as a system. Often, however, only one cause of losses is investigated, without regard to other concurrent hidden causes. Such partial restoration or correction may adversely affect the balance of the system and actually make the

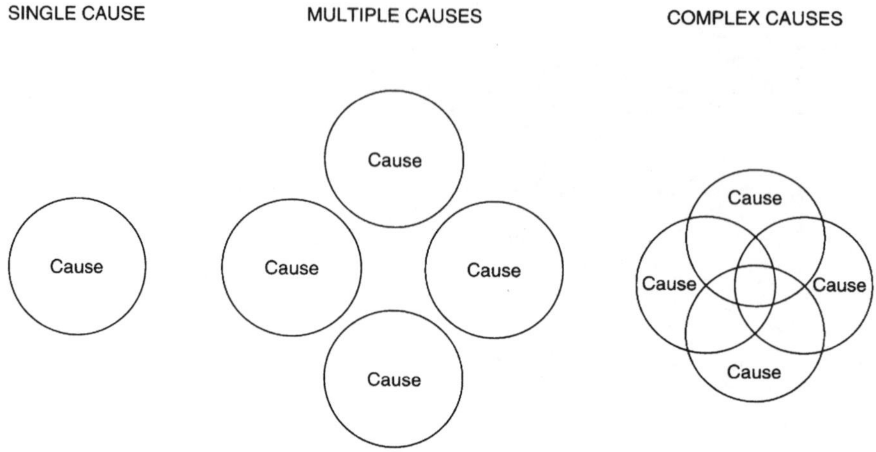

Figure 1.4 The causes of losses.

original problem worse. Losses, especially when resulting from complex causes, are never resolved by means of a traditional, problem-solving, "What are the causes of losses?" approach.

There are many troubles with equipment and quality for which no good solution was found in spite of various countermeasures taken in the past. Some production managers and maintenance engineers believe that there is no solution other than the replacement of old with new machinery. The alternative for overcoming such difficulties is to make, first of all, a careful examination of all pertinent phenomena. It is important to check precisely all available data, such as the occurrence and trend of malfunctions, the progress of deterioration, design conditions, operating circumstances, and atmospheric conditions. A project team, comprised of appropriate personnel from the production, maintenance, and plant engineering departments, is the most effective approach to finding solutions.

1.4.2 Equipment defects grow[2]

States of equipment that cause losses are referred to as equipment defects, and are classified as follows in accordance with their magnitude and based on their influence on losses.

Major defect. A single defect in a piece of equipment that can cause its breakdowns and operation stoppages (function-loss breakdowns). In terms of causes of losses, it is usually classified as a "single cause."

Medium defect. A single defect in a piece of equipment that can reduce its function (function-reduction breakdowns), but allows for continuous operation. In terms of causes of losses, it is usually classified as one of "multiple causes."

Minor defect. A single defect in a piece of equipment cannot cause losses in itself. Only when a particular combination of such single defects occurs by chance do these defects result in losses. In terms of causes of losses, it is usually classified as resulting from "complex causes."

The traditional common-sense approach to factory management never envisioned that the minor defects defined here cause breakdowns—for example, scant dirt, abrasion, distortion, play, looseness, or scratches in equipment. It is very rare, however, for a single minor defect to eventually result in losses. Such troubles, furthermore, are never resolved by a conventional approach.

From the TPM standpoint, minor defects are regarded as the probable cause of losses, unless these defects are absolutely determined to be completely unrelated to losses. Therefore, minor defects must be thoroughly eliminated without deliberation as to whether they contribute to losses. If minor defects are ignored, based on an overly optimistic view, they grow into medium and then major defects. That progression constitutes a deterioration of equipment as caricatured in Fig. 1.5.

Except for cases in which the causes are clearly proved—for example, various mistakes in stress calculation, material selection, welding and installation through equipment design, fabrication, and construc-

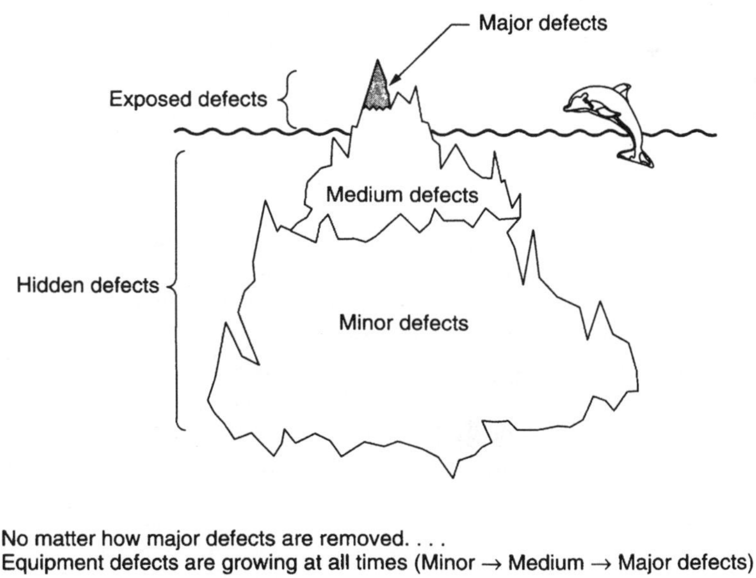

Figure 1.5 Equipment defects grow.

tion periods—it is not an exaggeration to state that almost all breakdowns occurring on a continuous basis are produced by medium and major defects which began as minor defects. In order to attain Zero Breakdowns, therefore, it becomes an absolute necessity to eliminate all defects whether they be minor, medium, or major.

Nevertheless, factory management, embracing traditional quality control concepts, criticizes this TPM approach with rejoinders such as, "It's overmaintenance!," "It's impossible to remove minor defects completely at a reasonable cost!," "It makes no sense to remove minor defects without estimating the impact on quality and breakdowns!," and so on. The seven-step program of operators' autonomous maintenance, which is based on concrete methodologies and technical foundations, is the response to such statements.

1.4.3 Forced deterioration and natural deterioration

Each component of a piece of equipment has its own lifetime and breakdown characteristics. Fig. 1.6 shows a composition of breakdown rates of component parts, or, in other words, the probability of the occurrence of equipment breakdowns in accordance with servicing time. The resulting chart is well-known as the bathtub curve and is divided into three periods as seen in the illustration.

Early breakdown period. All parts of equipment installed on the shopfloor, particularly mass-produced parts, are presumed to be highly reliable. In reality, however, a great number of early breakdowns due to engineering mistakes occur during start-up and commissioning in many plants. Plant engineers often make mistakes either because they underestimate the allowance for parts' strength (a result of an inadequate understanding of kinetic operating conditions), or because they fail to respect the design conditions and instructions given by vendors.

Occurrences of these early breakdowns vary widely in accordance with the structure, component parts, fabrication, installation, and operating conditions of equipment. Generally, occurrences are less frequent when qualified, mass-produced equipment is installed. When equipment is purchased according to the user's own specifications, or is designed and fabricated by the user, the breakdown rate is higher.

Breakdowns are purposely hastened to occur as early as possible by means of test runs during fabrication or after installation, and during the initial start-up operation of the entire plant. In other words, this commissioning period is the time to shift plant conditions from an early breakdown to a chance breakdown period. In TPM, it is known as "preventive engineering" whereby every effort is made through engi-

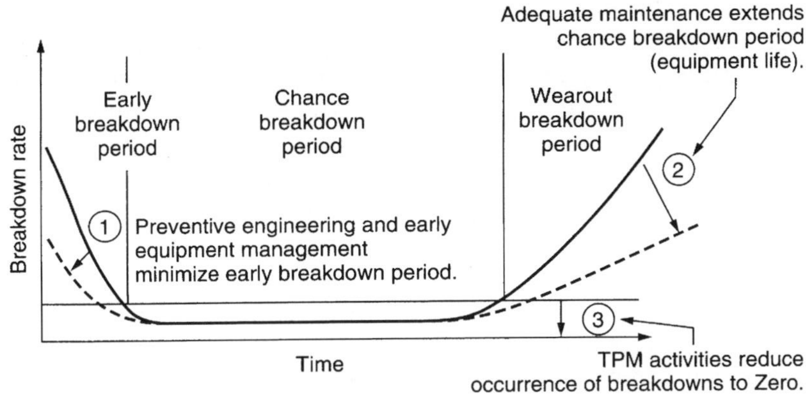

Figure 1.6 The bathtub curve and TPM activities.

neering, purchasing, construction, and commissioning to minimize or prevent troubles so commercial production may begin as early as possible. Further details in this regard are discussed in Chap. 2.

Chance breakdown period. If all malfunctions are remedied, plant operations are successfully initiated and stabilized as a system. The breakdown rate thereafter becomes nearly regular. Equipment and its component parts break down by chance. Unusually frequent breakdowns are caused by neglect of proper operation and maintenance rather than by inherent equipment weakness due to poor engineering and construction which were not corrected during the commissioning period.

Wearout breakdown period. Every component of a piece of equipment breaks as the end of its life expectancy approaches. The breakdown rate of equipment as a system increases at this time. A critical breakdown of this type terminates the service of equipment. If, however, parts are replaced by timely maintenance, the life of equipment can be extended during the chance breakdown period insofar as it may be cost effective to do so.

The deterioration of component parts causes equipment to break down. Deterioration of equipment is divided into two types: natural deterioration and forced deterioration.

As time passes, equipment is continuously worn according to the deterioration characteristics of each component part, even though the equipment was properly designed, fabricated, installed, operated, and maintained. This type of deterioration is referred to as natural deterioration. Equipment life in this case is labeled as inherent life.

Forced deterioration is defined as equipment that deteriorates more rapidly due to human performance rather than inherent characteristics. Examples are partially overloading particular parts by poor design, lack of cleaning and lubricating, and undertightening of bolts. More radical cases are mistakes in operating, manipulating, and repairing equipment. As a matter of fact, most breakdowns recurring daily in many factories is caused by this forced deterioration.

1.4.4 Hidden defects and exposed defects

As discussed in connection with losses, distinguishing between hidden and exposed defects helps to gain a better understanding of equipment defects which, otherwise, is not achieved.

Hidden defects are invisible for certain physical reasons.

- Equipment is contaminated with dust, dirt, or any other foreign substance and, as a result, defects are invisible. Ultimately, when equipment gets grossly contaminated, nobody even gets close enough for adequate inspection.
- Critical areas of equipment are not visible by way of ordinary human posture because of poor configuration of equipment.
- An excessive safety cover is installed. Its removal and installation require too much time and effort. As a result, the cover is not removed often for inspection.
- Equipment is designed without careful consideration for ease of inspection and cleaning. Neither an inspection port nor an entry hole is installed.

Or, defects remain undetected due to a lack of proper human vision or attitude.

- The importance of visible defects is underestimated on the basis of traditional thinking.
- Visible defects are not recognized as potential causes of losses because of the lack of proper education.

On the other hand, those defects that anyone readily recognizes are called exposed defects. In the presence of proper education, along with a certain amount of common sense and technical skill, even most hidden defects are recognized at a glance. With these points in mind, several simple prescriptions are suggested.

- Expose the hidden defects.
- Deliberately stop or interrupt equipment operations for inspection prior to the occurrence of breakdowns.

- Restore equipment defects promptly and completely.

With this kind of regimen, breakdowns never occur.

Autonomous maintenance is the answer to the question as to what and how operators must perform in keeping with the prescriptions listed above. Prior to a more detailed discussion, a review, in Chap. 2, of what has been said so far about TPM is in order.

Notes

1. Adapted from Soiichi Nakajima, *TPM Development Program* (Productivity Press, 1989), p. 4.
2. Skill management concepts developed by Masakatsu Nakaigawa, such as the importance of cleaning and minor defects, and the measurement of equipment effectiveness, which contributed to the creation of the basic TPM concepts.

Chapter

2

A Summary of TPM

2.1 Six Major TPM Activities

Today's TPM was born on the shopfloor of Japanese manufacturers during the late 1970s and has been making progress since then, basically as methodology for plant maintenance. As an increasing number of factories have implemented TPM concepts, a large body of experience and wisdom has been accumulated. Recently, various types of company-wide or factory-wide programs of this type bearing the title of TPM have been developed and applied even to administrative departments.

TPM, in the strict sense, refers to small group activities calling for total employee involvement and implemented primarily by the production, maintenance, and plant engineering departments to maximize productivity. In other words, it is a strategy adopted by all personnel who are involved directly with manufacturing to realize Zero Accidents, Zero Defects, and Zero Breakdowns.

It is, of course, necessary to match the development of TPM with the conditions that exist in each company or factory, such as manufactured goods, plant configuration, organization, local history, and culture at the plant site. In general, however, TPM consists of these six major activities, summarized in Fig. 2.1.

1. *Elimination of six big losses* based on project teams organized by the production, maintenance, and plant engineering departments
2. *Planned maintenance* carried out by the maintenance department
3. *Autonomous maintenance* carried out by the production department
4. *Preventive engineering* carried out mainly by the plant engineering department

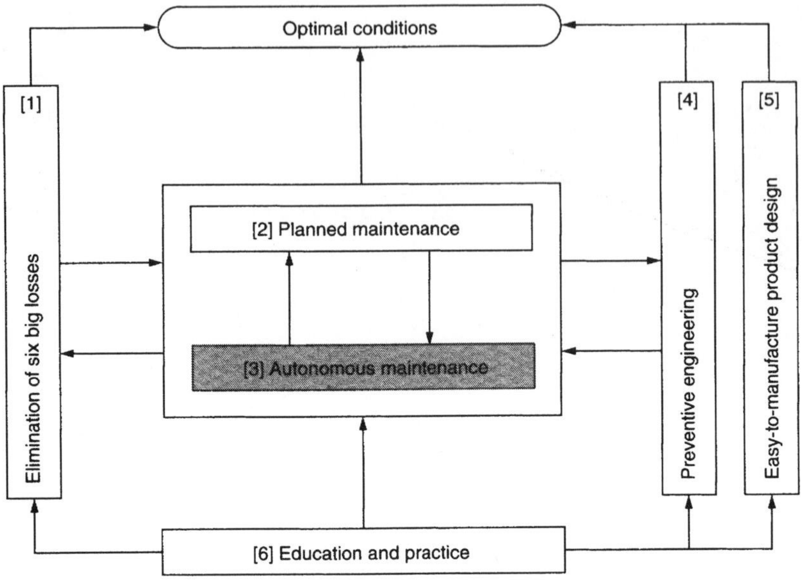

Figure 2.1 Six major TPM activities.

5. *Easy-to-manufacture product design* carried out mainly by the product design department
6. *Education* to support the above activities

TPM can be successful in achieving significant results only with universal cooperation among all constituents involved with the six activities listed above and illustrated in Fig. 2.2. In the absence of any one of these elements, satisfactory results cannot be expected.

Once a decision has been made to initiate TPM, company and factory leadership should promote all six of these activities despite excuses that may come from various quarters. The pursuit of short-term improvement in productivity by way of misinterpreted TPM principles should also be avoided.

2.1.1 The elimination of six big losses

Most of the six big losses in the past have been either overlooked or approached by traditional problem-solving techniques. Sometimes, personnel are frightened by huge losses exposed or projected through the calculation of overall equipment effectiveness. As an effective response to these problems, project teams are organized with appropriate membership drawn from the production, maintenance, and plant engineering departments. The activities of these teams are often directed by production personnel and consist of two categories.

Problem elimination. Current problems are eliminated in a relatively short time of three to six months. This approach helps to eliminate troubles on the shopfloor and, therefore, to reduce operators' anticipated workload during future autonomous maintenance activity. It also demonstrates to operators that persistent chronic losses can be reduced to Zero.

Innovative approach. Prior to the implementation of TPM, only traditional remedies based on established concepts were carried out. Most plant engineers and maintenance personnel have no experience in taking measures toward the reduction of the six big losses to Zero. Sometimes, they also do not believe that the achievement of Zero Breakdowns and Zero Defects is possible. In practice, these goals are attainable through innovative approaches based on Zero-oriented concepts. Success achieved in a limited area of process can be extended to the entire plant.

2.1.2 Planned maintenance

A planned maintenance system is established mainly by a maintenance department. Few, if any, desirable results can be expected, however, unless maintenance personnel make a concerted effort quickly. Such a response is required to set the stage for the mainline activities implementing the TPM system on the shopfloor. These can be divided into four phases.

1. Reduce variability of parts life
2. Extend parts life
3. Restore deteriorated parts periodically
4. Predict parts life

In some companies, these four phases are subdivided into a seven-step program for establishing planned maintenance which coincides with operators' activities. After parts life has been maximized through these efforts, a periodic maintenance system can be established. For critical equipment, parts life is dealt with by monitoring its status and replacing the related component parts prior to breakage. In this way, a predictive (or condition-based) maintenance can be applied in order to extend equipment life to its utmost.

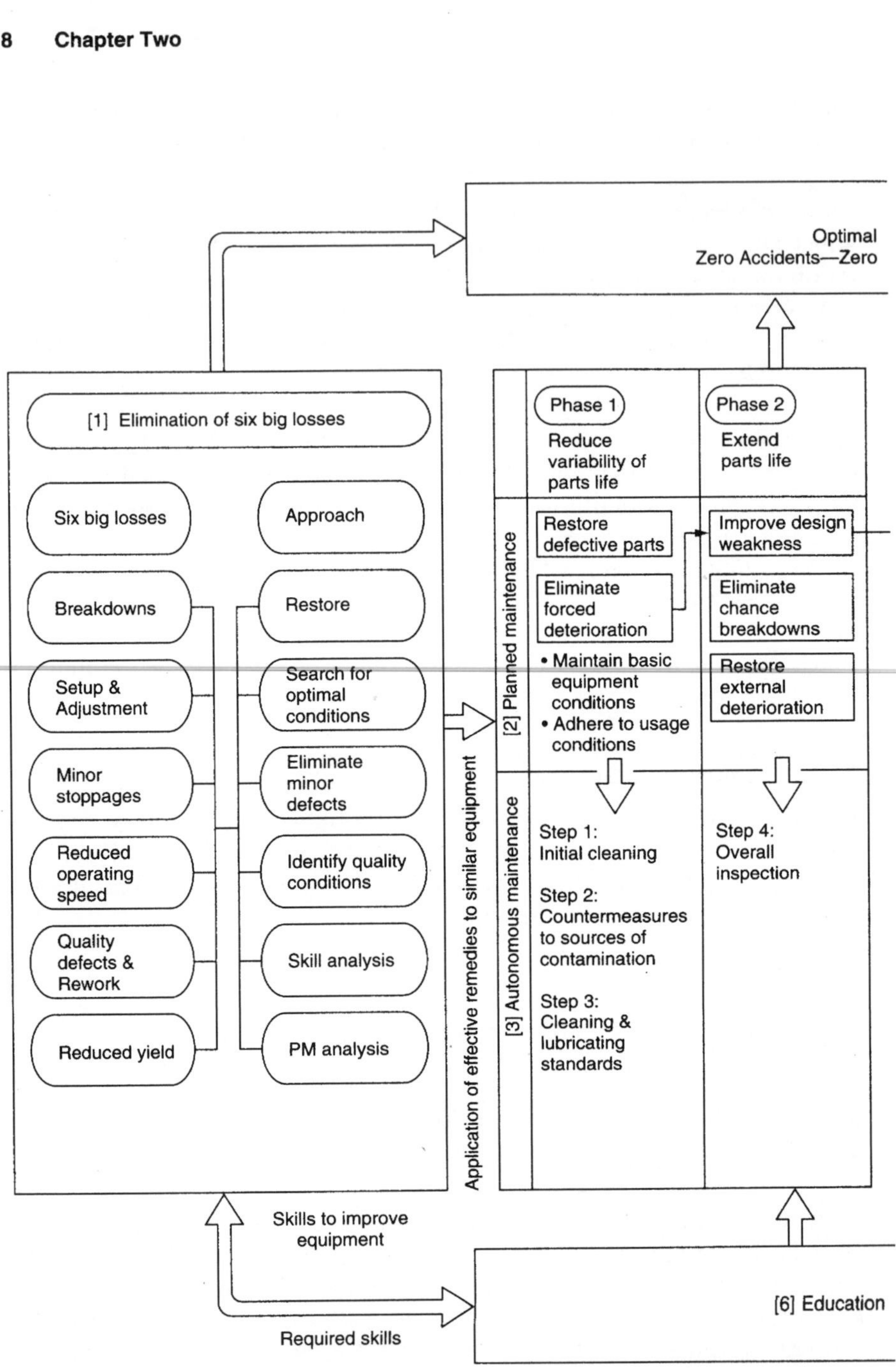

Figure 2.2 An outline of the TPM system.[1]

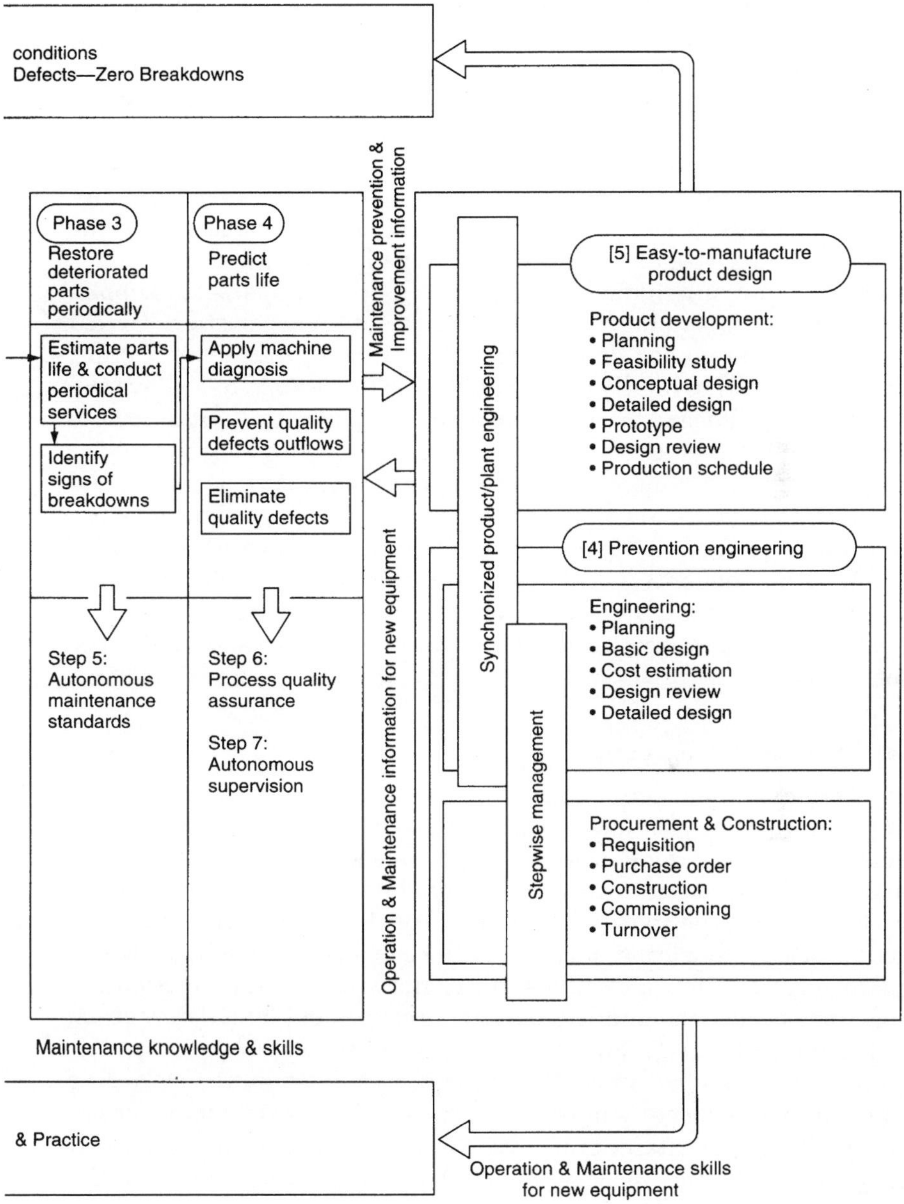

Figure 2.2 *(Continued)*

2.1.3 Autonomous maintenance

Autonomous maintenance activities are carried out by operators with the technical assistance of maintenance personnel. Operators are trained in the seven-step program (step-by-step education and practice) to achieve these major objectives: establish the basic equipment conditions (cleaning, lubrication, and tightening); observe usage conditions of equipment; restore deteriorated parts through overall inspection; develop into a knowledgeable operator; conduct autonomously supervised operator's routine maintenance. These fundamental maintenance activities performed by operators, following rules set by the operators themselves, are called autonomous maintenance.

Autonomous maintenance is programmed so that operators supplement activities to establish the planned maintenance system carried out by the maintenance department. The work of these two groups, therefore, must be carefully coordinated. It is impossible to achieve the first of the four phases, "to reduce variability of parts life," without close collaboration between the production and maintenance departments.

To establish an autonomous maintenance system, at least three and sometimes four years are needed. Although it is a very time-consuming, factory-wide activity involving all employees will result in significant effects and benefits. On the other hand, activities to eliminate the six big losses can bring definite results in a shorter period of time. In the latter case, however, benefits are limited to a small area of the plant.

2.1.4 Preventive engineering

Many problems that occur during the commissioning period for starting up a new production line must be resolved. "Early equipment management" entails a sequence of well-managed corrective actions to transform operations into commercial production as soon as possible. For this purpose, the causes of troubles must be eliminated not only in this commissioning period, but also in the earlier period, when a series of plant engineering tasks takes place. Instances include conceptual design, basic design, detailed design, procurement and fabrication, installation, test runs, commissioning, and turnover.

Moreover, a trade-off in conflicting equipment attributes regarding function and other factors must be examined. This review may encompass reliability, maintainability, economy, operability, and safety. It may provide for the discovery of solutions to problems and the improvement of future equipment on the basis of past experiences. This kind of effort is referred to as preventive engineering and includes all of the preventive activities taken in each stage, from plant engineering, procurement, and construction, to turnover as a whole.

2.1.5 Easy-to-manufacture product design

Despite the shortening life cycle of goods, the diversified needs of consumers must be satisfied in terms of product attractiveness, design, quality, and price so that competitiveness may be maintained in the worldwide marketplace. As a result, these challenges in manufacturing, which were once difficult to solve with only the shopfloor's efforts, can now more easily be eliminated if ease in manufacturing and quality assurance are built in at the product design stage. Not enough experience, however, has been accumulated yet in this area, and it will continue to occupy an increasingly important position in TPM.

2.1.6 Education

The activities just mentioned are not to be carried out by external TPM specialists. Although current internal capabilities may not be sufficient to achieve the goals of TPM, all participating employees must proceed with activities by themselves to master the necessary knowledge and skill by taking every available opportunity for education in these regards. No aspect of TPM can ever be achieved without suitable education as a foundation. In other words, the operator's education in the seven-step program of autonomous maintenance is an important contribution to the successful implementation of a TPM system in the production department.

2.2 The Effects and an Evaluation of TPM

TPM is not an end in itself. It is only a means to attain the managerial goals of the company or factory. Companies must primarily earn profits and improve employees' quality of working life (QWL). From TPM's viewpoint, it means that manufacturers must strive to improve productivity (P), quality (Q), cost (C), delivery or inventory (D), safety (S), and morale (M), sometimes abbreviated as PQCDSM.

The basic policy of TPM, as described in Table 2.1, is applied by incorporating it into the midterm managerial plans of a company. More concrete TPM issues can be identified by analyzing the kinds of managerial issues that must be tackled, along with the reasons and the purposes for doing so. Of course, every level of organization, such as the company, factory, department, section, or team, has different functions, missions, authorities, and responsibilities. TPM issues, therefore, differ accordingly. By making analyses from the highest level to the lowest, all employees must determine TPM issues and targets for themselves.

It is most essential, however, to reach a consensus by way of thorough discussion before launching TPM activities based on total

TABLE 2.1 TPM Basic Policy

Attain high quality and low production cost by means of all employees' efforts to overcome tough competition in the world marketplace
• Improve reliability and maintainability of equipment so as to realize high quality and productivity. • Foster development of knowledgeable operators. • Reinforce company's competitiveness by making the best efforts to develop new products and accelerate their development speed. • Increase value added per each employee, develop new production technologies, and promote automation. • Realize safe and favorable workplace by way of total employee involvement.

employee involvement. In many companies, leaders, frontline managers, and engineers have frequent meetings. The mutual understanding and consensus that result have important implications for future TPM development.

Actual targets of TPM are fixed more concretely in terms of PQCDSM. For example, initial conditions of 65 percent overall equipment effectiveness, 1000 breakdowns per month, 10 accidents per year, and only 1.8 suggestions per person-year can be improved over a four year time period to incidences of higher than 90 percent effectiveness, fewer than 10 breakdowns, Zero accidents, and 50 suggestions. The indexes that measure the effects of activities are assessed during the preparatory stage of TPM implementation. The conditions represented by these indexes are called "bench marks" and are used as criteria to estimate the progress made by TPM in the future.

Table 2.2 and Fig. 2.3 show, respectively, examples of targets and effects of TPM in a middle scale automobile parts manufacturer which employs approximately 1000 workers. It should be noted that actual and successful results occurred over a four-year period beyond the bench marks measured when TPM activities were initiated.

2.3 The TPM Master Plan

How must these effective activities be developed? In the preparatory stage, coordinated primarily by the TPM office, a three- to four-year plan of execution is prepared. It is referred to as the "TPM master plan." Figure 2.4 shows an example of a master plan carried out by a diesel engine division with 900 employees. While TPM was being developed there, plant facilities also were moved to a new site.

The schedule for proceeding with the six major TPM activities must be carefully set up, taking technical levels and current plant conditions into account. In the event that the number and expertise of engineers are not sufficient to proceed with all of these activities simultaneously, appropriate priorities may be assigned.

TABLE 2.2 TPM Targets

▪ Quality defects (Scrap + Trial run + Rework)	Reduce to 1/4 of bench mark
▪ Defective rate during commissioning period	0.1% (3 months after start-up)
▪ Productivity	Enhance 50% higher than bench mark (150%)
▪ Total equipment effectiveness	Improve 15% higher than bench mark (85%)
▪ Occurrence of sporadic breakdowns	Reduce to 1/50 of bench mark
▪ Workforce reduction by means of automation	100 employees
▪ Number of suggestions	6 Nos/month-employee
▪ Commissioning period reduction	Reduce 30% of bench mark
▪ Business expansion thanks to new products	30 million dollars
▪ Industrial accidents	0
▪ Productivity of administrative departments	Enhance 50% of benchmark (150%)

For example, before starting overall shopfloor activities, managers and engineers work for a half year toward the elimination of the six big losses. The most troublesome pieces of equipment are chosen as pilot models. By thoroughly eliminating breakdowns and quality defects, frontline leadership improves its technical skills and reduces the actual load of the shopfloor. Operators then initiate their autonomous maintenance program. The maintenance department concurrently develops activities to establish a planned maintenance system in an organized way by following along and helping with the operators' efforts.

In some companies, a preparatory period of several months to a half year, called Step 0, is spent cleaning and tidying up the shopfloor, and working on the elimination of the six big losses. Promoting safety and proceeding with introductory education for frontline personnel, in response to poor prior operating and equipment performance, are other important activities.

2.4 The TPM Steering Organization

To implement TPM successfully, a steering organization fills an important role. Its key points are described in this next section.

2.4.1 An overlapped small group organization

In accordance with the managerial structure in all departments, small groups, usually with five to seven people, are organized to include all employees from top managers to shopfloor personnel. An intermediate

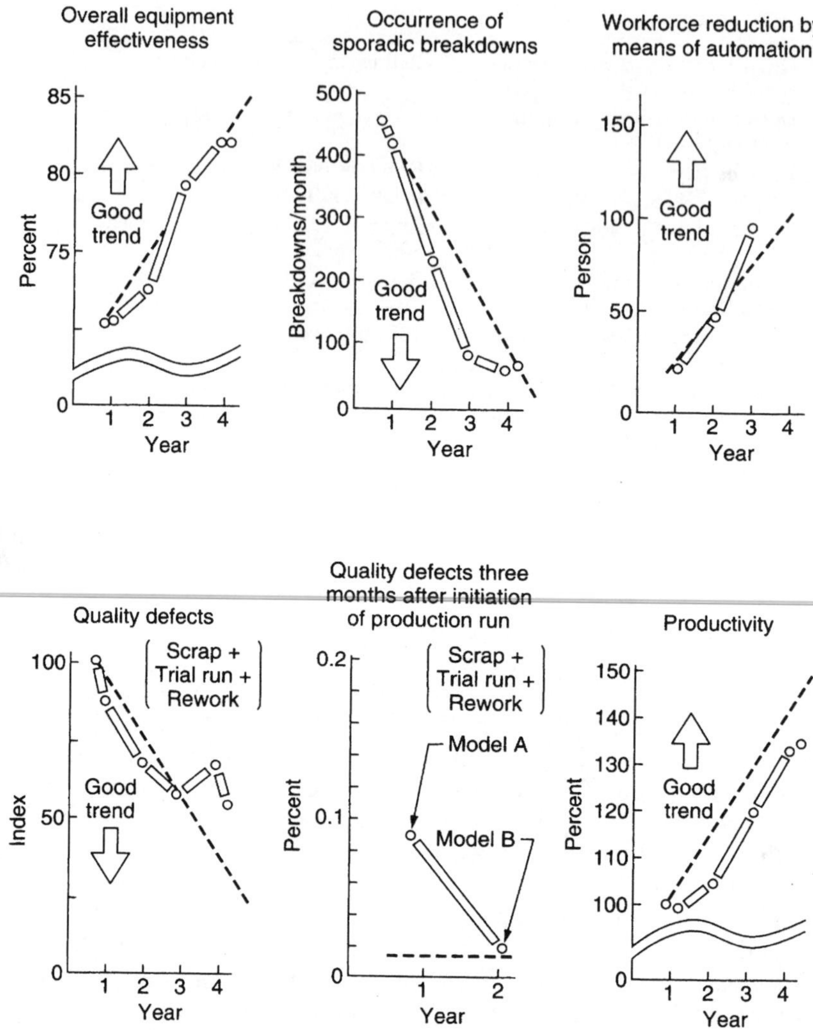

Figure 2.3 The effects of TPM activities.

group consisting of lower level leaders has the role of connecting those adjacent levels as a linking pin. Such a TPM steering structure is labelled an "overlapped small group organization." The relevant terms frequently referred to in this book are delineated in Fig. 2.5.

In the production department, these small groups consist of operators who develop autonomous maintenance and are simply called PM groups. Higher level groups consist of supervisors or team leaders led by a section manager, and are referred to as the managers' groups. The equipment handled by them is called the managers' model.

2.4.2 The committee and the project team

Apart from the overlapped small group organization, various other types of committees are often formed. According to past experiences with TPM implementation at many companies, it is, however, better to minimize such bodies. Too many independent task groups may obscure the existence of the essential overlapped small group organization. Moreover, people tend to lose focus when they belong to many different groups. It is always important to develop TPM activities in the format of small groups and to organize special committees only when absolutely necessary.

On the other hand, project teams consisting of personnel chosen from the production, maintenance, and plant engineering departments are very effective in addressing particular issues. Accordingly, expertise and skills from different technical areas are utilized to resolve problems arising from the shopfloor and, in a broader context, to eliminate the six big losses. These teams also help to promote understanding among departments through better communication.

Whenever planned targets are achieved, the project team must critically review the program in terms of its results and effects, and then identify countermeasures adequate to prevent a recurrence of the problems. After adequate reports are submitted to leadership and related departments, the special project teams should be dissolved promptly.

2.4.3 The TPM office

It is not an exaggeration to state that the success of TPM depends first on the strong determination of top management and second on the manager of the TPM office. Because this administrative manager plays a very important role in TPM, he or she must be selected from those who have sufficient managerial and engineering expertise, along with capabilities in planning, organizing, and leading. Office staffs must consist of several outstanding engineers who report directly to top management.

The TPM office is responsible for the total planning that encompasses all tactics related to TPM activities, and it also assists other concerned departments. It, therefore has many different tasks exclusively related to autonomous maintenance. In actual operations, the office implements some operations, but only plans others to be carried out by the department in charge.

Preparation period

- Take the lead in understanding the nature of TPM by participating in seminars or plant visits.

- Help in the preparation of the master plan (basic policy, background of implementation, major issues, steering organization, schedule, etc.) made at each level and unit of managerial organization such as the company, division, factory, department, and section.
- Estimate the cost and budget in terms of personnel, equipment, subcontracting, in- and out-of-house education and any other supporting expenditures.
- Negotiate with labor unions and arrive at an agreement.
- Establish bench marks, indexes, and methods to evaluate status quo and progress. Help with assessment in these regards.
- Plan and conduct the introductory education.
- Plan and conduct a campaign to let all employees become generally familiar with the concept of TPM.
- Plan a TPM steering organization.
- Conduct, generally at factory level, a kickoff rally in which top management announces the launching of TPM in the presence of all employees and other involved parties in and out of the company.

Autonomous maintenance

- Edit and issue TPM newsletters periodically.
- Edit and report major measuring indexes such as statistics on breakdowns, quality defects, and overall equipment effectiveness.
- Plan and conduct events such as TPM activity conferences, contests, and lecture meetings.
- Edit and report on the actual performance of PM groups in terms of a summary of activities, the frequency of meetings, the total time spent, and any other pertinent information.
- Participate in various meetings to assess the actual and general progress of TPM activities, and to assist PM groups.
- Receive suggestions from the shopfloor, arrange for their examination, and publish the results of the review. Edit and report these summaries and the number of suggestions.
- Edit and report statistics on the four lists which cover defective areas, questions, sources of contamination, and difficult work areas (see Chap. 5).
- Advise and teach concepts on how to develop TPM.
- Plan and conduct stepwise education.
- Arrange for the autonomous maintenance audit.

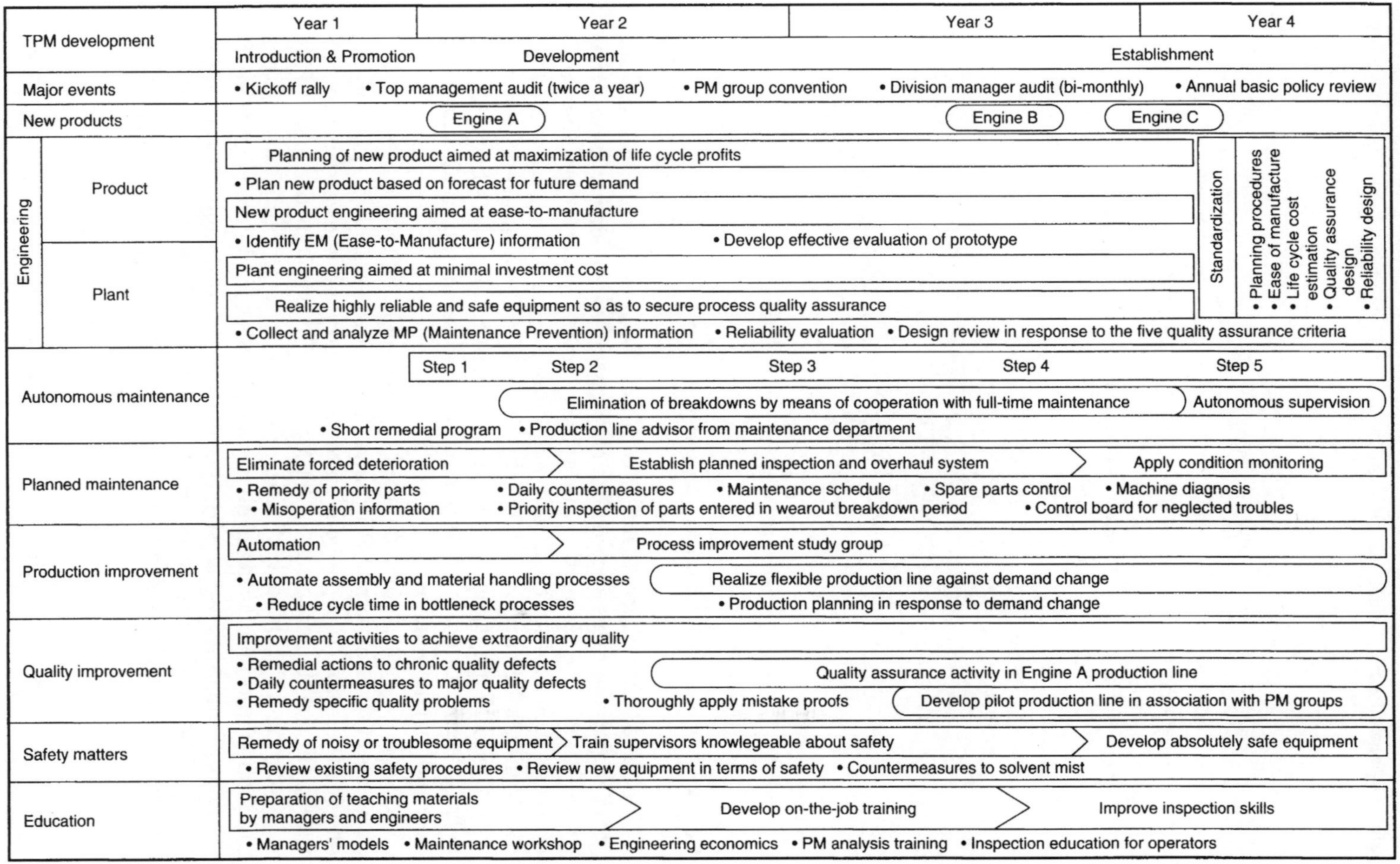

Figure 2.4 The TPM master plan.

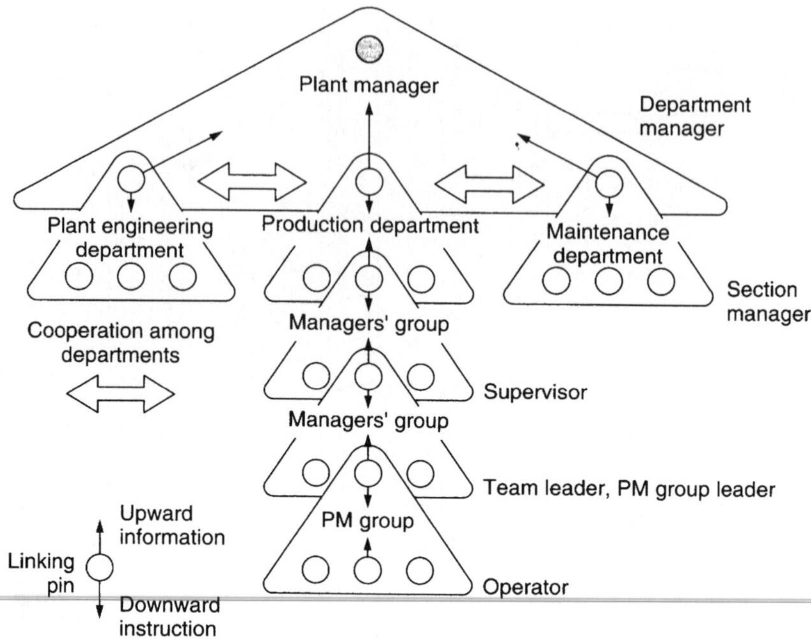

Figure 2.5 An overlapped small group organization.

The TPM office supports not only autonomous maintenance, but alsc other major activities such as the elimination of the six big losses, the establishment of planned maintenance, and preventive engineering systems. It is essential that any negotiations and adjustments among departments be carried out by the department managers in concert with the TPM manager. The TPM office continually helps to sustain a definite commitment among personnel by disseminating all information and instructions regarding TPM.

The three or four years required to implement TPM may seem to be an excessively long time. This time, however, passes very quickly. If, on the other hand, people hold on to customary methods and past attitudes, the implementation of TPM certainly will fail because TPM involves the difficult task of realigning the roles and functions of the production and maintenance departments. Steadfast determination of company leadership along with the planning and administrative capability of the TPM office is required to overcome these obstacles and to assure the realization of optimal conditions in a manufacturing plant.

Note

1. Adapted from Soiichi Nakajima, *TPM Development Program* (Productivity Press, 1989), pp. 18–19.

Chapter

3

The Five Countermeasures to Achieve Zero Breakdowns

3.1 The Basic Strategy to Attain Zero Breakdowns

The basic strategy to achieve Zero Breakdowns is to "expose the hidden defects, and to deliberately interrupt operations prior to the occurrence of breakdowns to remedy equipment defects promptly." In other words, all barriers that obscure equipment from adequate human scrutiny must be removed. To this end, all of those personnel concerned with the manufacturing business need to change their conventional attitudes or ways of thinking.

As examples, the following countermeasures are considered in the situations which are discussed on page 12 of Chap. 1.

- Remove contaminants stuck to the surface of equipment to make defective areas visible.
- Change the shape, attachment, and location of the equipment's component parts to make essential areas visible from normal human posture and positions.
- Modify safety covers for easier removal and reinstallation.
- Install inspection ports or entry holes in appropriate locations.
- Inspect for deterioration by condition monitoring and carrying out timely overhauls.
- Attach great importance to minor defects by abandoning traditional attitudes.
- Foster technical skill through adequate education to detect equipment defects.

Almost no effective results can be expected, however, if these countermeasures are taken at random. By organizing them into the following five countermeasures against the essential causes of breakdowns (as illustrated in Fig. 3.1), and then by the thorough and systematic execution of each, Zero Breakdowns are achieved.

1. Establish the basic equipment conditions (cleaning, lubrication, and tightening).
2. Adhere to the usage conditions of equipment.
3. Restore deteriorated parts.
4. Correct design weakness.
5. Enhance operating and maintenance skills.

These five countermeasures are discussed in more detail in the next sections.

3.1.1 Establish basic equipment conditions

These three—cleaning, lubrication, and tightening—are known as basic equipment conditions. To attain Zero Breakdowns, the basic equipment conditions must first be established by thorough cleaning,

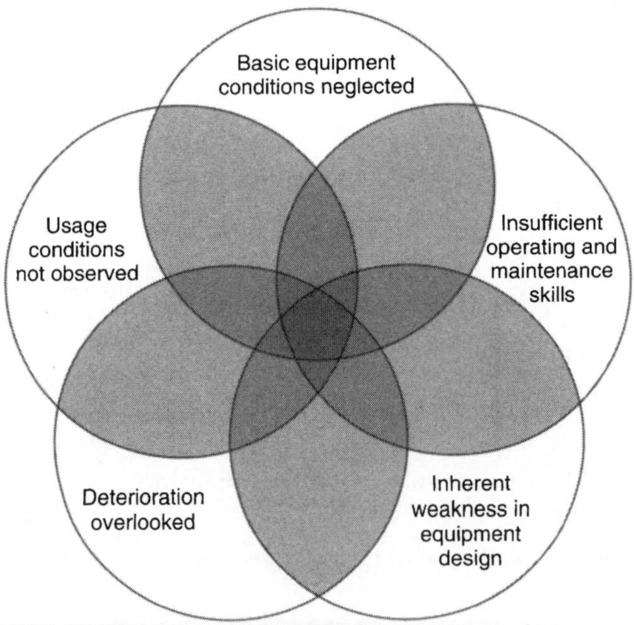

Figure 3.1 The five essential causes of breakdowns.

lubrication, and tightening—measures that were frequently neglected over a long time in the past. Preventive measures are also taken to easily maintain these established conditions by improving equipment and work methods.

Cleaning. Cleaning is the elimination of any foreign substances stuck to equipment and its surroundings. Those substances, which may result in abnormal vibration, seizure, uneven abrasion, burnout, defective actuation, lower accuracy, and so on, cause minor stoppages and breakdowns in sliding parts, hydraulic, pneumatic, electrical, and instrumental systems, and cause corrosion, leakage, and plugging in static equipment such as vessels, heat exchangers, and piping systems. Equipment deterioration resulting from insufficient cleaning is a typical example of forced deterioration. In addition, foreign substances adhering to or mixed into raw materials and workpieces can result in quality defects.

The cleaning discussed here is neither a matter of keeping the factory clean in terms of superficial good appearance, nor is it traditional industrial housekeeping. It is, on the other hand, cleaning to establish basic equipment conditions and to expose hidden defects as a means of attaining Zero Breakdowns and Zero Defects.

Therefore, any portions of equipment covered with safety guards and remaining invisible must be completely exposed and cleaned. By looking into and touching every nook and corner, frontline personnel can discover various types of defects such as abrasion, play, looseness, notch, deformation, leakage, overheating, vibration, and noise. In other words, the hidden defects can be exposed.

In some companies where cleaning is understood in a superficial sense, people are often inclined to strive only for visual cleanliness, usually by painting equipment and buildings. These actions hide defects for the sake of appearance and, therefore, should never be taken to the exclusion of more significant measures.

From the TPM viewpoint, the purpose of cleaning is to expose hidden defects so as to be able to correct them before they lead into medium or major defects. On the other hand, through thorough cleaning, operators can learn about the structure and function of equipment. In other words, cleaning is, in part, education that uses equipment as subject matter and the shopfloor as a classroom.

Lubrication. The many and various parts to be lubricated include rotating and reciprocating parts that are built into equipment. Unless lubricants are properly supplied to these parts to provide and maintain sufficient lubrication, abnormal operating conditions, such as vibration, noise, overheating, abrasion, and seizure, will occur and will sometimes result in breakdowns. Lubrication is one of the most important prerequisites for preventing deterioration and maintaining the reliability of equipment.

These exposed losses are easily recognized and are much less significant than hidden losses, which seem to have no apparent relationship to insufficient lubrication, such as the recurrence of a large number of quality defects, or difficulties in setup and adjustment. As an example, a plant could cut five percent of its electrical consumption merely by implementing a thorough lubrication control program. Such a typical case may suggest how poor lubrication overloaded equipment and, as a result, how improper power transmission badly influenced quality and accelerated deterioration.

Upon closer examination, too many of these poor lubricating conditions are found. For instance, there is often no lubricant in air lubricators or reservoirs. Lubricating points and supply pipes are covered with dirt, and excess grease contaminates equipment. In an extreme case, a large quantity of oil leaks through a weld crack at the bottom of a reservoir. Sometimes, lubricants are not supplied to lubricating surfaces because of clogged oil supply tubes.

Generally speaking, the importance of lubrication is understood by everyone. Nevertheless, serious discussion and effective countermeasures are frequently neglected, even though the troubles caused by poor lubrication continually occur. The fact that insufficient lubrication does not always result in immediate breakdowns or quality defects tends to cause the need to be overlooked during a busy routine. These situations, therefore, can serve as the best and most typical examples of hidden defects in many companies. The specific reasons why proper lubrication is played down so frequently are as follows:

- Frontline personnel are not taught about lubrication and its importance in preventing actual losses.
- Even managers and engineers have insufficient appreciation for the importance of lubrication.
- Lubricating standards (lubricating points and surfaces, type of lubricants, quantities, intervals, and tools) are not prepared.
- Education and practice regarding lubrication have not taken place.
- Too many types of lubricants must be applied at a large number of lubricating points.
- Not enough time is allowed for lubrication.
- There are too many difficult and time-consuming lubricating points at which to apply lubricants.

The plant engineering department, being in a hurry to hand over equipment to the production and maintenance departments, prepares lubrication standards by simply combining vendors' operation manuals without careful study. On the shopfloor, frontline managers and engineers are always pressed to resolve current breakdowns and quality defects, and, as a result, do not have enough time either to remedy equipment or to revise given standards in order to facilitate lubrication under actual operating conditions.

Maintenance personnel also are very busy with too many work orders for repair. Operators, too, are always attending to changeovers and a large number of minor stoppages which leave them with insufficient time to follow given lubricating standards. Everyone hopes that no severe breakdowns occur before the ends of their shifts. This situation is typical of the vicious circle which can be observed in many factories.

Moreover, the number of lubricating points is skyrocketing in accordance with the increase of automation in the repetitive manual work shop. If no serious response is made, the burden on operators and maintenance technicians will increase yearly, and the situation cannot help but get worse.

Tightening. Almost all equipment is put together with fasteners such as nuts and bolts. Damaged, loose, or lost fastening parts cause vibration, misalignment, or defective actuation in equipment, and result in minor stoppages and breakdowns. It is not rare for a plant handling highly flammable materials to have serious accidents involving fires or explosions. These are often consequences of leaked flammable agents caused by loose bolts at flanges.

When one of several bolts is loose, it does not necessarily result in immediate leakage. Repeated expansion and contraction as a result of temperature change or vibration can allow one loose bolt to loosen other bolts. The resulting play then becomes larger and larger until it eventually exceeds a limit in tolerance. A minor defect, such as one loose bolt, if neglected, grows into a moderate and finally a major defect, which sometimes results in a sudden breakdown, or, at worst, catastrophic failure.

According to a very detailed survey on the causes of breakdowns made in one plant, 60 percent of them involved, to some extent, defective nuts and bolts. In another plant, an overall inspection revealed even more astonishing results: that is, 1091 out of 2273 bolts were loose or lost, for an amazing 48 percent incidence of defects. In addition, various troubles frequently occur due to over-, under-, or uneven tightening of nuts and bolts on die and jig assemblies. Improper tightening also occurs due to poor skill in tooling during changeover.

3.1.2 Adhere to the usage conditions of equipment

A piece of equipment and its component parts are designed to function under given conditions of use within certain tolerances. For example, there are oil temperature, pressure, quantity, oxidation, and contamination with foreign particles in hydraulic systems; atmospheric temperature, humidity, dust, vibration, shock, and corrosive gas in instrumental systems; plus installation methods and locations, shape of actuator, and contact with workpiece in sensors.

Plant engineers frequently do not take into account conditions of use during the design stage. In the event of construction of a new plant or revamping of existing facilities, most of the machinery, in general, is purchased through vendors. Engineers also design some of their own specialized equipment: jigs, tools, material handling, and other auxiliary facilities. In these designs, small parts, such as sensors, air filters, regulators, lubricators, and valves, are often neglected and left to the detailed design of vendors or subcontractors.

If a plant is started up without careful consideration of the above points during the engineering and construction stages, frequent minor stoppages will occur afterwards. Sometimes, workpieces collide with and break sensors. In spite of the high reliability and long life of these mass-produced parts, lack of attention from plant engineers to such matters causes many problems throughout the future commercial operation stage.

Furthermore, no matter how correctly plant engineers design and install equipment, accurate information on usage conditions, such as plant engineers' design concepts and vendors' detailed instructions, may not be properly submitted to the production and maintenance departments when the plant is handed over. In addition, operators, due to poor education, cannot faithfully observe usage conditions. The shopfloor personnel, in the worst case, are completely unaware of the existence of such conditions. These situations inadvertently cause various types of forced deterioration of equipment. In other words, equipment is gradually being broken all the time due to improper human actions instead of natural wear.

3.1.3 Restore deteriorated parts

There are some companies that are proud of the number of suggestions received from the shopfloor and over react by trying to improve everything. Although it certainly is important to respond to ideas coming from the front line, suggestions that lack a focus or a definite objective must be sorted out. Otherwise, remedial actions based on vague ideas can make matters worse.

A piece of equipment, in general, is designed on the basis of a balance in strength and accuracy in its component parts. The restoration of equipment is the recovery of this balance. Therefore, a simple idea proposed by those who have little technical knowledge cannot be expected to produce effective results. More particularly, equipment purchased on the market has almost no room for improvement by amateurs.

When a breakdown does occur, attempts to remedy the problem often are of a symptomatic nature. In other cases, only the broken parts are repaired or replaced without sufficient follow-up on the possible deterioration of related parts. Since such deteriorated parts do not always break, this kind of partial improvement or restoration is frequently inadequate.

Suppose, for instance, that a shaft breaks at a notch. Prior to simple replacement with a spare part or design change to reinforce it, it is more important to investigate, first of all, whether usage conditions were closely adhered to and whether overall balance of equipment continues to exist. After a detailed survey of nearby related parts has been made, the shaft must be replaced. Only when such a survey concludes, however, that a design change is absolutely necessary, will required action be taken to alter the shape, material, or thermal treatment of the existing shaft.

In an actual case in a factory, the real cause of the breakage did not exist in the shaft. It was, to the contrary, excessive vibration resulting from abrasion in a keyway that produced the break at the weakest notched portion of the shaft.

Many companies are eager to encourage any improvement ("Kaizen") focused on equipment and work methods. In spite of successful progress at the beginning, it usually fades after a while. Because there are not many situations that can be handled by those who have little technical expertise, few remarkable results can be anticipated. In other words, such attitudes reflect slipshod managerial attitudes and a lack of leadership in expecting productive suggestions from frontline personnel who have not received adequate education.

When the basic equipment conditions are well established and usage conditions are simply followed, a considerable number of breakdowns will be eliminated, as illustrated in Fig. 4.2 in Chap. 4. There must be expected, nevertheless, a certain number of design or installation mistakes resulting from technical demands facing engineers in charge of plant engineering and construction. Most of these relatively early mistakes can be resolved easily through test runs and commissioning.

When a breakdown occurs in a piece of equipment that already has entered into the chance breakdown period, partial restoration or improvement approached only in terms of broken parts, and without proper attention to deterioration in related parts will lead to similar breakdowns in the same equipment. When deterioration in adjoining parts is neglected and remains hidden, it is safe to say that the related parts must have deteriorated when a certain part was broken. In most cases, the deterioration barely remained within safe tolerance and was simply overlooked.

As time passes, equipment normally deteriorates gradually. The most deteriorated or weakened parts are broken by one another. Before deciding on piecemeal improvement, one should return to the original drawings and specifications to recognize all possible instances of deterioration. Although this regimen, which restores overall balance in equipment, may seem to be lengthy, it is, in reality, the shortest path to Zero Breakdowns.

3.1.4 Correct design weaknesses

Generally speaking, manufacturers of equipment do not have enough experience in fabricating goods with equipment that they have manufactured themselves. In spite of their best efforts to respond to the users' requests, equipment all too frequently is designed based on engineers' imagination rather than on experience with past problems in its operations.

Even if equipment is designed in-house, most plant engineers have limited experience and know-how in production and maintenance. Because personnel who design equipment and those who use it have different backgrounds and belong to different departments, problems are destined to occur at their interface. Correcting weakness in design, therefore, becomes necessary during the commercial production stage, and must usually be initiated by shopfloor personnel.

To some, this conclusion may seem to conflict with assertions made in prior discussions. The question to be addressed, however, is not one of improvement for its own sake, but one of the policy and methods to achieve such improvement. Techniques for effective improvement constitute one of the most important resources for companies that are competing using similar machinery purchased in the same marketplace. These companies will, on the other hand, never become competitive by applying strategies that lack clearly stated goals and objectives.

In each step of autonomous maintenance programs, frontline personnel focus on specific pinpointed areas, such as sources of contamination, difficult work areas, lubrication, fastener, power transmission, and so on. After clearly setting out intentions and targets, they concurrently tackle equipment malfunctions and design weaknesses. In particular, they remedy breakdowns and quality defects in accordance with technical and logical inferences based on the assessment of current machine working conditions.

In this way, shopfloor people can come up with excellent ideas if they are properly educated on how to accomplish improvement with the suitable technical assistance of specialists. It should be noted that most of the case studies introduced in this book describe typical actions taken by operators under such circumstances.

3.1.5 Enhance operating and maintenance skills

Every company employs many different countermeasures against breakdowns. This is, unfortunately, a problem with personnel who are inclined to focus on hardware such as equipment, jigs, tools, dies, raw materials, products, and so on. To the contrary, less obvious "software," such as technical knowledge and skills of employees on the production front line, merit less attention. Regretfully, the term software often has implied to managers a bureaucratic control system involving monotonous standardization or dull documentation which burdens operators and maintenance technicians with impractical standards and work procedures. Then, too, many managers continually complain that the shopfloor personnel do not follow the rules. This attitude reveals their lack of managerial capability.

Any operator or maintenance technician should not desire purposely to misoperate and break equipment or to make mistakes in repairing it. They certainly would not prefer to work in a dirty factory or to be injured by an accident. Conversely, anyone directly involved in the manufacturing business actually wishes to work in an environment of Zero Accidents, Zero Defects, and Zero Breakdowns. Despite these natural preferences, they cannot achieve these goals in the absence of proper knowledge and skills.

Accordingly, many companies apply on-the-job methods for operator training. This kind of training, however, can be quite ineffective. It often results in nothing being done because too much reliance is placed on the shopfloor. There are only a few companies that provide a skill development program for individual employees, for example, by implementing a skill evaluation system, as shown in Table 3.1. Furthermore, a detailed study shows that there are many incorrect methods of equipment operation, maintenance, and inspection which were viewed as correct.

Therefore, a different approach in which techniques applicable to specific types of machinery or processes are identified and applied, is indicated. Subsequently, operators and maintenance personnel learn how to apply these particular operation and maintenance skills. In the meantime, it must be remembered that these essential techniques are always changing, especially in accordance with progress in automation technology.

TABLE 3.1 Operators Evaluate Knowledge and Skills by Themselves

Skill Evaluation Sheet

Process: ______ Self Evaluation No: ______
PM Group: ______ Date Evaluated: ______
Operator: ______ Group Leader: ______
Supervisor: ______

Definition	Level A (Junior instructor)	Level B	Level C (Average)	Level D	Level E
● Knowledge	Teach fellow operators based on pertinent knowledge	Widely apply proper knowledge to routine tasks	Properly apply indispensable knowledge to routine tasks	Necessary knowledge not fully learned yet	Sufficient knowledge not learned yet
▲ Skill	Train fellow operators based on superior skills	Widely utilize proper skills to routine tasks	Properly utilize indispensable skills in routine tasks	Necessary skills not fully acquired yet	Sufficient skills not acquired yet (experience less than one year)
Category					
L: Lubrication	● Teach general matters of lubrication	● Comprehend type of lubricants and lubricating methods ● Revise lubricating standards ● Analyze past problems caused by poor lubrication	● Comprehend type, characteristics and usage of lubricants ● Comprehend function and necessity of lubrication ● Comprehend lubricating points and surfaces ● Execute color lubrication control ● Know past breakdowns caused by poor lubrication	● Insufficiently comprehend type, characteristics and usage of lubricants	● Insufficiently comprehend lubrication

TABLE 3.1 *(Continued)*

	▲ Analyze and remedy breakdowns caused by poor lubrication	▲ Detect and remedy defective parts during routine inspection ▲ Suggest better and easier lubricating ▲ Manage local lubricants storage	▲ Properly lubricate following lubricating standards ▲ Inspect lubricating points and surfaces, and take necessary corrective actions ▲ Properly manage lubricants and tools	▲ Properly lubricate in a longer time frame	▲ Lubricate per request
J: Fastener	● Teach general matters of fasteners, tightening tools and methods	● Recognize fastener-related problematic area in process ● Revise fastener inspecting standards ● Comprehend past problems caused by poor tightening	● Comprehend mechanism of fasteners ● Comprehend type, name and characteristics of fasteners ● Recognize white and yellow match marks	● Insufficiently comprehend type, name, and characteristics of fasteners	● Insufficiently comprehend fasteners
	▲ Improve fasteners to reduce inspection time ▲ Improve equipment by analyzing problems caused by fasteners	▲ Partially improve equipment to remedy problems	▲ Properly tighten nuts and bolts using specified tools ▲ Inspect and tighten nuts and bolts following standards ▲ Inspect nuts and bolts, and take necessary corrective action ▲ Properly manage fasteners and tools	▲ Tighten nuts and bolts using specified tools	▲ Improperly tighten nuts and bolts at times

Details of educational methods in autonomous maintenance are discussed in Chaps. 4 and 8. Very successful case studies are described, and they are applicable not only to production and maintenance, but also to those departments indirectly involved in manufacturing, such as plant engineering, quality assurance, and any other administrative departments. Therefore, it becomes essential for managers and staff to thoroughly understand these educational methods and to apply them in routine management.

3.2 Who Takes the Five Countermeasures?

The five countermeasures discussed previously should never be implemented solely by the maintenance department, even though it is involved to some extent with all matters pertaining to equipment. If a company insists, nevertheless, on doing so, the number of maintenance personnel must be markedly increased. Otherwise, frontline personnel will be overloaded and will be forced to give only the appearance of implementing the five countermeasures. A large number of unreasonable requests and rules have, for a long time, been forced on the departments which are directly responsible for matters of manufacturing. Such a history, in many companies, has led to very poor plant conditions against which the five countermeasures definitively need to be taken.

It is apparent that manufacturing cannot be continued for long without the functions of both equipment operation and maintenance. These two functions, however, generally have been separated from each other. To successfully execute the five countermeasures, the functions and roles of the production and maintenance departments should be viewed, as illustrated in Fig. 3.2.

Accordingly, the production department takes charge of basic routine maintenance in addition to its conventional tasks relating to production. These supplemental efforts by operators are referred to as autonomous maintenance and consist of these activities:

1. Maintaining basic equipment conditions
2. Adhering to usage conditions of equipment
3. Visually inspecting for external deterioration of equipment to detect signs of abnormalities by the five senses and by relatively easy machine diagnostic techniques
4. Enhancing operating, setup, adjustment, and inspection skills

On the other hand, it is the maintenance department that takes charge of more sophisticated maintenance. These maintenance personnel's efforts are referred to as full-time maintenance and consist of the following activities.

1. Technically assisting autonomous maintenance
2. Definitely restoring defective parts in equipment by inspection, overhaul, condition monitoring, and machine diagnosis
3. Identifying appropriate usage conditions of equipment and taking necessary remedial actions when design weaknesses are discovered
4. Enhancing maintenance knowledge and skills

When the production and maintenance departments fulfill these designated roles, proper equipment conditions can be established and maintained by the approach of the five countermeasures.

3.3 Restructuring the Roles of the Production and Maintenance Departments

Until recently, most Japanese manufacturers, influenced by American-style factory management, clearly separated the roles of the production and maintenance departments. They were convinced that this style was the most effective way to utilize human resources. Operators concentrated on production with little knowledge about the structure and function of equipment. Concurrently, maintenance personnel took an indifferent attitude to work orders. As a result, operators and maintenance personnel went their own ways instead of following the path of mutual cooperation.

To keep up with recent progress in automation of assembly industries, various types of machinery are being installed in factories at a rapidly increasing rate. Shopfloor personnel, on the contrary, are gradually decreasing in number and are becoming older in average age. A more serious problem, widely found in many highly industrialized countries, is the tendency of the younger generation to lose interest in the manufacturing business. It is quite natural for young men and women who can choose their own future not to look forward to hard work in a dirty factory with dangerous equipment and a relatively low income.

In view of these developments, it becomes critical to the survival of companies competing in the worldwide marketplace to improve dramatically, over the next decade, their productivity and quality as well as the working conditions of the shopfloor. In this regard, TPM has rapidly become popular because companies can succeed in overcoming today's difficulties by restructuring the roles of the production and maintenance departments in accordance with the seven steps of the autonomous maintenance program.

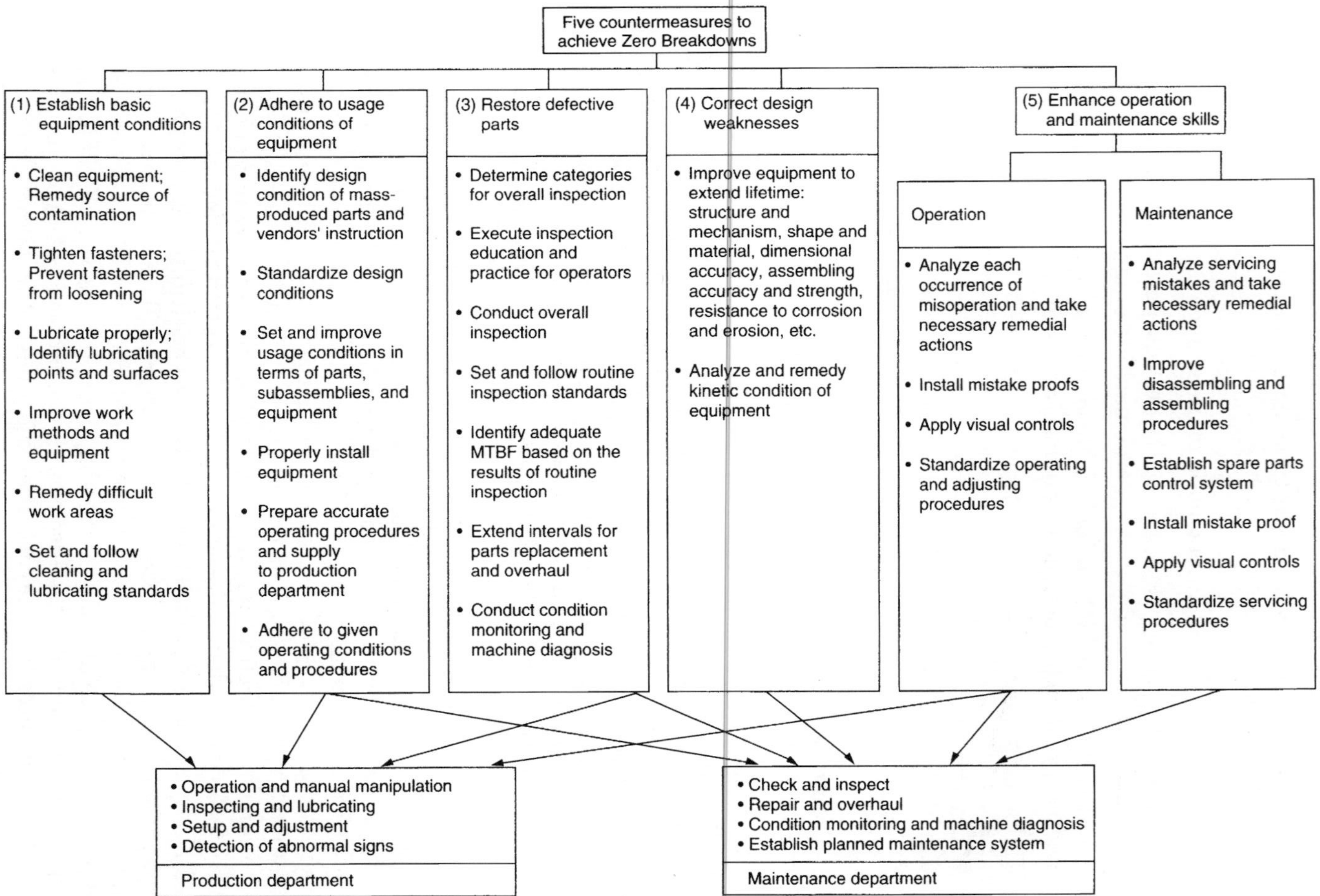

Figure 3.2 The five countermeasures to achieve Zero Breakdowns.

3.4 Allocating Roles to the Production and Maintenance Departments

Mutual cooperation among the production, maintenance, and plant engineering departments is the key to attaining Zero Breakdowns through the five countermeasures. On the contrary, poor communication and cooperation among these three departments causes enormous losses, even today.

It is apparent that the manufacturing business cannot exist without both production and maintenance functions involving the plant engineers' assistance. Only by synchronous and reciprocal efforts of these departments can equipment be efficiently operated and maintained. With today's increasing reliance on equipment, operators must take charge of some part of maintenance. This new role for operators is an essential ingredient of total maintenance. That is, the prevention of deterioration in equipment. Full-time maintenance cannot achieve its proper role until operators patiently establish and maintain the basic equipment conditions through simple, but minute procedures. The roles of the production and maintenance departments are delineated in Fig. 3.3.

To attain completely the aims of maintenance for maximizing equipment effectiveness, two types of activities must be performed simultaneously.

Maintenance activities restore and prevent equipment deterioration and breakdowns.

- Preventive maintenance:
  - Routine maintenance: Maintain the basic equipment conditions and replace deteriorating parts.
  - Periodic (or time-based) maintenance: Periodically inspect equipment and restore defective parts.
  - Predictive (or condition-based) maintenance: Correct equipment deterioration by condition monitoring and machine diagnosis.
- Breakdown maintenance: Correct equipment deterioration after the occurrence of breakdowns. This becomes unnecessary when TPM is successfully implemented.

Remedial activities extend equipment life, reduce the time spent for maintenance work and, eventually, make maintenance itself unnecessary.

- Reliability improvement: Prevent breakdowns by improving reliability of equipment.
- Maintainability improvement: Make maintenance work easier.

All of the above activities can be classified into three major maintenance activities. They are restoration, inspection, and prevention of equipment deterioration. Maintenance goals cannot be attained if any of the above activities is neglected, even though these activities differ from each other in terms of methods and effectiveness.

3.4.1 Maintenance activities for the production department (autonomous maintenance)

The production department focuses on the prevention of deterioration and takes charge of the following:

1. *Restoring deteriorated parts*
   - Establish and maintain basic equipment conditions (cleaning, lubrication, tightening).
   - Make proper adjustments of equipment during operation or changeover.
   - Record data on breakdowns, minor stoppages, and quality defects.
   - Collaborate with maintenance personnel to repair or improve equipment.
2. *Inspecting deterioration* (sensory observation by the five senses and easy machine diagnosis)
   - Perform routine inspection.
   - Perform parts of periodic inspection.
3. *Preventing deterioration*
   - Make minor repairs (easy part replacement and makeshift measures)
   - Report promptly and accurately to the maintenance department occurrences of breakdowns or quality defects.
   - Assist maintenance personnel in repairing equipment.

Among these activities, maintenance of basic equipment conditions and routine inspection are the most essential. Conducting them exclusively by the maintenance department, however, results in an excessive work volume. It might be feasible if additional maintenance personnel are mobilized as required, but to do so is usually impossible.

If, on the other hand, operators deal with deterioration, more certain execution and effective results are expected. Operators, obviously, are always working around their equipment and should

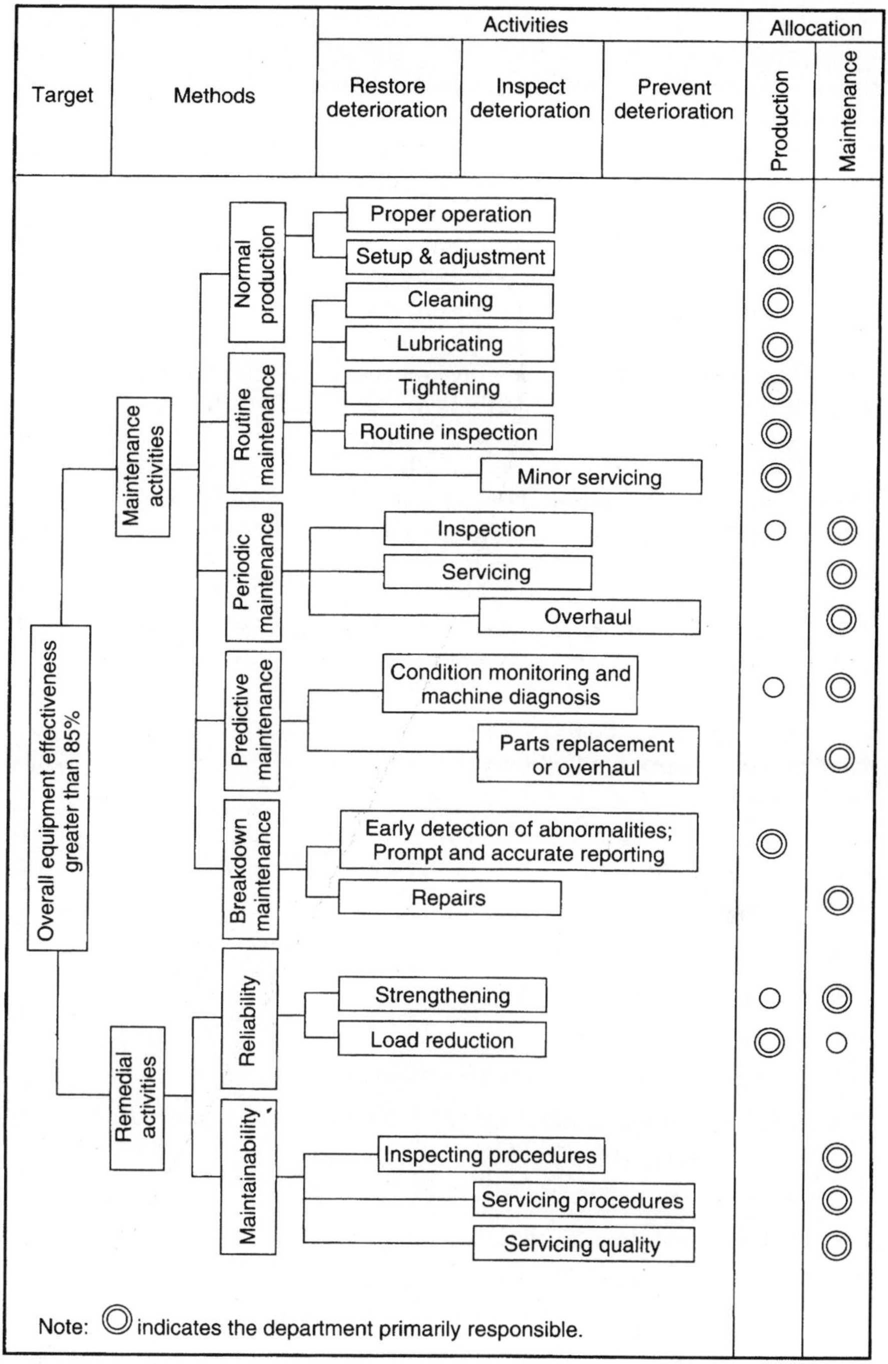

Figure 3.3 The roles of the production and maintenance departments.[1]

know best about daily equipment conditions and their trends. This approach is not at all comparable to the situation in which maintenance personnel come running to make repairs only at the occurrence of sudden breakdowns.

However, bringing operators to the point of conducting autonomous maintenance activities requires continuous education and practice for as long as three or four years. Furthermore, a satisfactory operators' maintenance system must be organized and developed in accordance with detailed concepts of the seven-step program. A company should not become frustrated with the absence of immediate results and definitely should not take shortcuts by, for example, neglecting the first half of the program in order to move on more quickly to the second half or quality matters. Any such efforts are certain to fail.

Autonomous maintenance depends mainly on knowledgeable operators who are able to assume responsibility for routine aspects of their work by themselves. When TPM is not fully recognized for what it is, neither Zero Breakdowns nor Zero Defects will be attained. All TPM activities, including autonomous maintenance, primarily enhance human capability and change ways of thinking rather than equipment. Recognizing this point, leadership must incorporate TPM into mid- and long-term managerial plans of the company or factory.

3.4.2 Maintenance activities for the maintenance department (full-time maintenance)

The maintenance department focuses on inspection and prevention of equipment deterioration, and it takes charge of the following as maintenance specialists:

1. *Restoring deteriorated parts:*
   - Repair sudden breakdowns.
2. *Inspecting deterioration:*
   - Conduct periodic inspections and overhauls.
   - Perform condition monitoring and machine diagnosis in order to estimate critical parts life.
3. *Preventing deterioration:*
   - Perform professional maintenance work.
   - Plan and manage maintenance work to be subcontracted.
   - Improve equipment reliability and maintainability.
   - Manage information regarding improvement and maintenance prevention and submit it to the engineering department.

Faithful restoration of deteriorated parts in equipment, which used to be unfeasible even for a maintenance department, is now carried out by operators. The maintenance department must, instead, be responsible for higher maintenance levels such as more sophisticated improvement, condition monitoring, machine diagnosis, risk management, and so on. It also must report to the plant engineering department more detailed information about design weakness discovered in existing equipment. Thanks to the reduced recurrence of breakdowns and extended parts life—accomplished with the operators' efforts—maintenance personnel can now conduct various routine tasks. As in the past, these were not realized in spite of their good intentions.

To this end, full-time maintenance activities must be developed in close cooperation with autonomous maintenance, as summarized here.

Establish a planned maintenance system. No matter how faithfully a maintenance schedule is followed, it does not mean that a planned maintenance system is well implemented. To arrive at truly planned maintenance, the four phases aimed at Zero Breakdowns must be carried out. After maximizing parts replacement intervals by means of extension of parts life, definite periodic or time-based maintenance is carried out. In reference to critical equipment, predictive or condition-based maintenance may then be applied. The maintenance system which can be realized in this approach can be considered truly planned maintenance.

Quickly and promptly treat work orders from autonomous maintenance. A huge number of deteriorated or defective parts in equipment will be discovered by operators through the autonomous maintenance program, particularly in the periods of Steps 1 and 4. Those restorations or minor modifications of equipment that cannot be provided by operators will be requested of the maintenance department. As a result, maintenance personnel will be frequently overloaded due to a flood of work orders from PM groups. Measures for preventing this situation should be carefully considered in advance. Slow responses can detract from operators' motivation and from their autonomous maintenance activities. Handling of such requests is expressed by the suitable index called the work order completion rate (the number of work orders completed divided by the number of work orders). This completion rate must be kept higher than 85 percent at all times.

Assist and guide autonomous maintenance. Without suitable assistance and guidance by full-time maintenance, operators cannot develop autonomous maintenance activity smoothly. Maintenance personnel must teach operators safety matters, proper cleaning, and usage of hand tools in Step 1, and provide detailed technical assistance in overall inspection practice programmed in Step 4. On the other hand, maintenance personnel frequently learn from operators' unsophisticated and spontaneous questions.

Analyze misoperation information and plan countermeasures. The maintenance and production departments must cooperate to analyze causes and to plan preventive measures in response to operators' mistakes. Mistake proofs or visual controls might be installed when necessary. To realize this kind of remedial action, misoperation-related information must be recorded early on. Of course, serious misoperations must be treated promptly at each occurrence. A more detailed discussion is provided in Chap. 5.

Improve technical levels of maintenance, and develop more sophisticated techniques. First, maintenance personnel should have sufficient capability to be able to teach operators. To eliminate the six big losses and to establish a planned maintenance system, there are many technical issues to be resolved. In addition, more sophisticated techniques are required to respond to today's technical progress in automation. It is also important to bring maintenance standards and relevant documents up to date in terms of contemporary progress in equipment.

Record maintenance information and evaluate results. No matter how detailed the information recorded in daily reports or equipment ledgers may be, it cannot be clear and distinct enough to serve as an effective guide to maintenance tasks. Therefore, information regarding type, location, and causes of breakdowns must be edited as clearly sorted records, trends, and Mean Time Between Failure (MTBF) data. The computer application in these areas, such as data keeping and effective evaluation of maintenance results, is one of the most essential prerequisites for establishing planned and predictive maintenance systems.

Report maintenance prevention information to the plant engineering department. No matter how considerable the efforts that have been made, improvement of an existing plant yields very small benefits when the plant's life cycle cost in terms of attributes, such as reliability, maintainability, operability, safety, and economy, are taken into account. Those characteristics are almost completely determined in the engineering, fabrication, and installation stages.

Therefore, operating, breakdown, and quality defects data, as well as improvement information regarding such characteristics, must be fed back to the plant engineering department to impact on the development and design of a future plant. This kind of information is referred to as maintenance prevention information, or MP information in short. It must be standardized and compiled into written documents which can readily be shared by anyone concerned.

In many companies, the maintenance department is frequently not interested in improvement of maintainability. In other words, no one can comprehend how slowly maintenance personnel repair or overhaul equipment. The eagerness of maintenance personnel to tackle work promptly and effectively is much poorer than that of operators to increase productivity. Maintenance department managers and engineers must give maximum attention to this point.

Note

1. Adapted from Soiichi Nakajima, *TPM Development Program* (Productivity Press, 1989), p. 47.

Chapter

4

The Autonomous Maintenance Program

4.1 The Aims of Autonomous Maintenance

Autonomous maintenance activity has two aims. From a human perspective, it fosters the development of knowledgeable operators in light of their newly defined role. From an equipment perspective, it establishes an orderly shopfloor where any departure from normal conditions may be detected at a glance.

4.1.1 Reconsidering the operator's role

In Japan, today's assembly industry is constantly accelerating the automation of repetitive manual work based on recent progress in computer and micro-electronic technologies. Automation technology is one of the most important resources for manufacturers' survival. Although various kinds of repetitive work are still found on the shopfloor, the trend of the times is definitively toward automation.

A sizable number of companies that lack the technical or financial capability to follow this trend transfer their plants overseas. In many cases, however, the easy approach of depending only on cheap labor is likely to fail. Their products may be competitive in price, but they are not in quality. Automation not only reduces the production cost, but also improves the quality and yield of products by eliminating errors caused by manual work.

Traditionally, two types of repetitive manual work have existed in the factory. One is sophisticated repetitive work such as manipulation of machine tools. The other is primitive repetitive work, such as workpiece setting onto machines, along with parts assembling, bolt tightening, materials handling, and cleaning.

The modern mass production industry, following the Ford production system, is based on technologies such as standardization of parts, division of labor in the assembly line, and simplification of human work. The industrial engineering concept, or Tailorism, has supported this development. In this environment, most workers have been required to repeat simple work all the time without any knowledge about the structure and function of either the equipment they are operating or the product they are manufacturing. In doing so, they are not expected to consider or judge anything during their routine work.

Now, however, in keeping with the progress in automation of manual work, workers at last are liberated from simple and monotonous repetitive work and engage in truly human work. On the other hand, the functions which can only be performed by human beings are indispensable for today's highly complex and sophisticated production systems. To recognize and respond to this fact, TPM uses the term "operator" instead of "worker."

For example, in an automobile parts manufacturing plant, an almost fully automated process was implemented. By using computer controls, nearly one hundred machines could be run with the assistance of only two operators. A large number of minor stoppages, however, occurred as metal chips frequently got twisted around tools and jigs. The operators, as a result, were always bustling around the process. It, however, may be more cost-effective in this kind of situation to increase the number of operators if productivity and depreciation are taken into account.

This case shows, therefore, that top management and factory leadership are likely to make mistakes in this regard as follows:

- They do not tackle seriously the kinetic operating conditions of the equipment.
- They do not know the importance of the operator's role and potential.
- They pay too much attention to computer, robotic, and other high-tech procedures, and neglect the basic and low-tech procedures.
- They believe that by eliminating a number of operators, production costs can be reduced correspondingly.

Unfortunately, traditional and conservative ideas regarding operators still exist within the leadership of many companies, even though the relationship between human beings and machinery is changing daily. Some managers may think it unnecessary to educate shopfloor personnel because production in a fully automated plant can be conducted simply by following the computer's instructions. These people always isolate manufacturing activities from total corporate operations as a result of their short and narrow vision. In other words, they continuously attempt to achieve the maximum output of production-related departments with minimum input.

To examine the manufacturers' business as a whole requires that all personnel concerned with manufacturing go beyond understanding the roles of the production and maintenance departments. They must now take into account all departments which directly or indirectly impact on routine production; not only the production and maintenance departments, but also the quality assurance, material handling, plant engineering, product design, administration, and any other supporting departments. Success will be evasive unless, by careful consideration, these same personnel provide for and achieve the optimal and most flexible integration of function and roles among all of these departments. What is most essential in the managerial organization and production system is a very dynamic and flexible response to drastic changes in society and technology during the course of the second industrial revolution.

TPM, with the participation of all employees, means reviewing the function and roles of all personnel and analyzing the interfaces of the various departments. One outcome of this process is the realization that the operator's role should be revised. As a result of the autonomous maintenance program, the thought patterns of frontline managers and operators will naturally change. The seven-step program is designed to facilitate this transition. This program is especially appropriate because higher-level management is generally less flexible than frontline personnel.

4.1.2 The knowledgeable operator

The term "knowledgeable operator" does not mean an operator who can fix equipment as well as be a maintenance technician. Rather, it emphasizes that an important aspect of an operator's skill is to detect signs of losses. That means the operator should be able to sense that something "funny" is going on whenever some unusual conditions exist during the operation of equipment and prior to the occurrence of breakdowns or quality defects.

Almost all losses, either breakdowns or quality defects, are preceded by some signs. Not only do minor defects occur, but so do certain other aberrations, such as unusual vibration, noise, odor, or overheating. Admittedly, before the implementation of TPM, it used to be impossible to detect such minute signs on a shopfloor flooded with excessive losses and abnormalities. Once the basic equipment conditions are nearly established by way of Step 3, the occurrence of breakdowns is generally reduced by half of the bench marks. It then becomes possible to detect certain signs.

Operators may, therefore, be expected to master the very early discovery of indications of abnormalities so as to prevent the occurrence of losses. Management needs to train operators who will be able either to report such situations to the maintenance department quickly and accurately, or to handle the problems by themselves immediately. To make this possible, operators must be, from the beginning of the autonomous maintenance program, educated about the basic structure and function of equipment, and later trained through practice on actual equipment.

This is not an easy task and requires a certain amount of time and money. It is, nevertheless, absolutely necessary in order to implement the TPM system across the entire workplace and, therefore, to achieve remarkable benefits.

4.1.3 Training operators toward a new type of engineering status

In the petroleum and chemical industries, which have a long history of automated process control, to prepare for emergencies operators are allocated to the control room in numbers that exceed the minimum needed for normal operations. In some advanced plants, operators are developing their technical skills by studying automated plant operations and emergency shutdown procedures using artificial intelligence computer technologies in redundant time to the exclusion of human intervention.

In the assembly industries, the aims are a little different, but some extra operators still need to be allocated to even highly automated processes so as to maintain a certain redundancy of human intervention. As reliance on equipment is increasing annually, risk and cost in plant investment also become higher. These trends result in a greater need to maintain flexibility in the production system by providing adequate redundancy between equipment and human beings, and by assuring a thorough understanding of kinetic operating conditions.

There is no one other than the operators who can understand kinetic operating conditions and accurately answer the questions posed by maintenance, plant engineering, and product design engineers because operators are working around the equipment all the time. If some of them are assigned as managers of the production department and leave the equipment, they will no longer be able to understand kinetic conditions.

Generally speaking, leadership tends to underestimate operators' hidden capability and potential. Accordingly, it has no feeling of guilt about keeping operators in monotonous repetitive work for their entire lives. If, however, suitable motivation, training, and opportunities are provided for operators, they attain remarkable ability which leads to highly desirable results.

Autonomous maintenance aims at fostering the development of these knowledgeable operators. In moving ahead, a company must train operators toward a new status of production engineers. This idea is now being tried in a few companies, and this development certainly can be expected to grow in the near future.

4.1.4 The orderly shopfloor

If a knowledgeable operator is reallocated to a traditional dirty plant where a huge number of minor stoppages and breakdowns are occurring all the time, he or she cannot utilize special capabilities at all. An "orderly shopfloor," or, in other words, a shopfloor where aberrations from normal conditions can be detected at a glance by anyone, is attained only when knowledgeable operators are manufacturing products in a process where optimal operating conditions are established.

This kind of process can be attained only when a necessary condition is met—the optimal status of both human beings and equipment. In addition, human beings and equipment must complement one another. Only under these circumstances will Zero Accidents, Zero Defects, and Zero Breakdowns be attained.

Optimal equipment conditions refer collectively to a status in which basic equipment conditions (cleaning, lubrication, and tightening) are firmly established, usage conditions of equipment are well observed, deteriorated parts are restored, and equipment is properly operated all the time.

To this end, the frontline managers must understand the aims of autonomous maintenance from the perspective of both human and equipment elements, as illustrated in Fig. 4.1. These managers also must make great efforts toward fostering knowledgeable operators, and thereby contribute significantly to the realization of an orderly shopfloor.

The visual control which makes the largest contribution to the above efforts is described here. Visual control refers to surveillance control that uses a device which can reveal at a glance the normal or abnormal status of physical and human behavior. For example:

- Place paint match marks on nuts and bolts to detect looseness.
- Put marks on gauges and indicate adequate limits of operating pressure and quantity of lubricant.
- Apply temperature sensing tape on motors and reduction gears to check for overheating.
- Paint designated color and arrow signs on pipes to show the type and pressure of fluid and direction of flow.

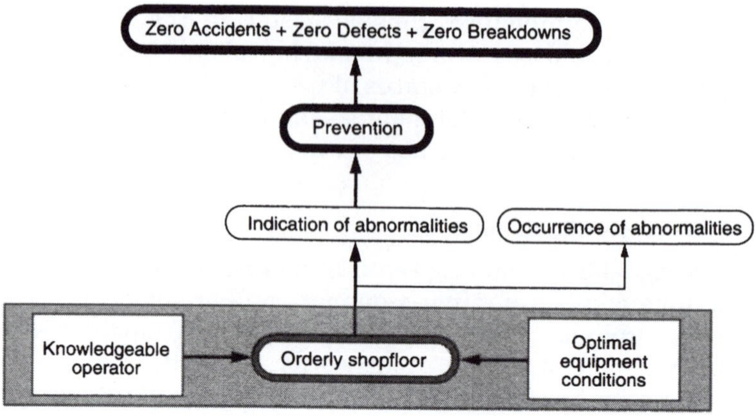

Figure 4.1 The orderly shopfloor and the Zero-oriented concept.

- Apply arrow signs and indicate rotating direction of belts, chains, and motors.
- Install inspection holes on safety covers or replace an existing cover with a transparent material.
- Install a small sign board and indicate the open/close status of a valve.
- Install sight glass to show the flow of cooling water, coolant, or any other fluid.
- Install a ruler to indicate the open/close positions of the duct damper and paint adequate range markings.
- Apply color lubrication control (see Chap. 7).
- Install a small streamer to indicate the air flow of blowers and ventilators.
- Assign and display code numbers for processes, equipment, major subassemblies, jigs, and tools.
- Assign and display code numbers for floors, shelves, and parts boxes.

In addition to the above, many gadgets have been devised by shopfloor personnel. Even though each of them taken by itself is only a minor idea, all of them applied collectively throughout the entire plant produce remarkable results. Thorough application of visual controls is one the most important requirements to achieve an orderly shopfloor. Of course, their application and handling procedures must be standardized. This allows all operators to detect any abnormalities at a glance and to respond immediately with conditioned, reflex-like remedial actions.

4.2 The Autonomous Maintenance Development Program

4.2.1 The seven steps of autonomous maintenance

In previous chapters, the natures of losses and countermeasures to eliminate them were discussed. In this section, a summary of the seven-step program arrives at a general view of the development of autonomous maintenance.

To this end, the causes of losses must initially be identified. Primarily, they are a matter of equipment breakdowns. Quality defects are also caused by problems with equipment, especially in an automated plant. Furthermore, equipment is designed, fabricated, installed, operated, and maintained by human beings. From this standpoint, it might be acknowledged that losses result from human thought patterns and behavior.

Therefore, without dramatic changes in our conventional way of thinking, Zero Breakdowns and Zero Defects can never be attained. The attitude most acutely required of today's manufacturers is an innovative approach that produces breakthrough countermeasures. The purpose of TPM is to achieve these ends, which result in the thorough improvement of existing equipment, and to effect changes in future equipment and products by applying experiences and lessons acquired along the way.

Zero Defects have existed traditionally only as a slogan in campaigns promoting quality. It is, nevertheless, impossible to achieve Zero Defects easily and assuredly in any plant. To paraphrase, one may ask, "Is there some good method applicable to any factory for succeeding regardless of present technical levels and expertise of frontline personnel, if everyone from top management to floor workers is sufficiently committed to this goal?" The answer to this question is the seven-step autonomous maintenance program, summarized in Table 4.1. Each step is sometimes divided into more detailed substeps.

4.2.2 The first stage of the development program (Steps 1, 2, and 3)

The first stage of the development program consists of Steps 1, 2, and 3. It is not only the starting point of autonomous maintenance, but also of all TPM activities. It focuses on creating the foundation of TPM by establishing proper cleaning, lubrication, and tightening of equipment.

The activities programmed in the first stage are the various remedial actions to restore deteriorated parts in equipment. The major objectives are: "Establish basic equipment conditions," and "Understand what autonomous supervision is."

TABLE 4.1 Summary of Autonomous Maintenance Program

Step	Major activities	Aims from equipment perspective	Aims from human perspective	Managers' supervision and support
1. Initial cleaning	▪ Thoroughly clean equipment and its surroundings ▪ Remove all unnecessary materials ▪ Write upcoming issues onto four lists	▪ Expose hidden defects by removing contaminants ▪ Restore defective areas in equipment ▪ Identify sources of contamination	▪ Become familiar with group activity by way of easy tasks such as cleaning ▪ Group leaders learn leadership ▪ Look at and touch every corner of equipment to enhance its care and to promote curiosity and questions ▪ Learn "Cleaning is inspection"	▪ Lead by staying one step ahead, comprehending TPM through practice, and demonstrating with examples of managers' models ▪ Teach defects of equipment ▪ Teach importance of cleaning, lubrication and tightening ▪ Teach "Cleaning is inspection"
2. Countermeasures to sources of contamination	▪ Remedy sources of contamination ▪ Prevent contaminants from irregular and undesirable dispersion ▪ Improve difficult cleaning areas to reduce cleaning times	▪ Prevent contaminants from generating and adhering to equipment in order to enhance reliability ▪ Definitely maintain equipment cleanliness so as to improve maintainability	▪ Learn motion and working mechanism of machinery ▪ Learn methods to improve equipment focused on sources of contamination ▪ Encourage interest and desire to improve equipment ▪ Feel pleasure and satisfaction with successful achievement of improvement	▪ Teach motion and working mechanism of machinery ▪ Teach where-where and why-why analyses to examine problem ▪ Assist in implementing ideas for improvement ▪ Promptly respond to work orders

TABLE 4.1 *(Continued)*

3. Cleaning and lubricating standards	▪ Conduct education for lubricating ▪ Develop overall lubrication inspection ▪ Establish lubrication control system ▪ Set cleaning and lubricating standards	▪ Correct difficult lubricating areas ▪ Apply visual controls ▪ Definitely maintain basic equipment conditions (cleaning, lubricating, tightening) to establish deterioration prevention system	▪ Set rules by oneself and follow them ▪ Know importance of following rules and autonomous supervision ▪ Encourage awareness of one's own roles and of teamwork	▪ Prepare lubrication control rules ▪ Provide education and practice in terms of lubrication ▪ Teach how to prepare cleaning and lubricating standards ▪ Assist actual preparation of standards
4. Overall inspection	By each inspection category: ▪ Conduct education and practice ▪ Develop overall inspection ▪ Remedy difficult inspection areas in equipment to reduce required time ▪ Set tentative inspecting standards	▪ Detect and remedy minute defects ▪ Thoroughly apply visual controls ▪ Improve difficult inspection areas ▪ Maintain established equipment conditions by means of routine inspection to improve reliability further	▪ Learn structure, function and inspection methods of equipment to master inspection skill ▪ Master easy servicing procedures ▪ Group leaders learn leadership through conducting roll-out education ▪ Learn recording, summarization and analysis of inspection data	▪ Prepare overall inspection schedule, check sheets, manuals, and other teaching materials ▪ Promptly respond to work orders ▪ Provide training for easy servicing ▪ Teach how to improve difficult inspection areas by applying visual controls thoroughly ▪ Teach inspection data handling

TABLE 4.1 Summary of Autonomous Maintenance Program *(Continued)*

Step	Major activities	Aims from equipment perspective	Aims from human perspective	Managers' supervision and support
5. Autonomous maintenance standards	▪ Set autonomous maintenance standards and schedule to finalize activities focused on equipment ▪ Faithfully conduct routine maintenance in accordance with standards ▪ Move forward aiming at Zero Breakdowns	▪ Assess successful remedies achieved in other processes, and apply them to similar equipment ▪ Totally review visual controls ▪ Preserve equipment in highly reliable condition along with operability and maintainability ▪ Realize an orderly shopfloor	▪ Understand equipment as a total system ▪ Develop ability to detect signs of abnormalities to prevent breakdowns ▪ Train knowledgeable operators ▪ Establish autonomous supervision system conducted by PM group	▪ Allocate inspection work between autonomous and full-time maintenance ▪ Teach basic maintenance skill and easy machine diagnosis ▪ Teach examples of breakdown prevention ▪ Teach particular function of each piece of equipment to understand equipment as a system
6. Process quality assurance	▪ Prevent outflow of defective products to downstream processes ▪ Prevent manufacturing of defective products ▪ Attain process quality assurance and move forward aiming at Zero Defects	▪ Assess process quality ▪ Attain a reliable process to prevent outflow of quality defects ▪ Assess quality conditions ▪ Attain a highly reliable process to prevent manufacturing of quality defects	▪ Train knowledgeable operators on equipment and quality aiming at new type of engineering status ▪ Attain autonomous supervision within each operator	▪ Teach quality specifications, quality causes and quality results along with their relationship ▪ Teach the five criteria for ease of observation ▪ Teach the five criteria for quality assurance ▪ Address matters of quality with cooperation by all related departments

TABLE 4.1 *(Continued)*

7. Autonomous supervision	▪ Maintain, improve and pass on current TPM levels	▪ Predict abnormalities to prevent breakdowns and quality defects prior to occurrence ▪ Attain Zero Accidents, Zero Defects and Zero Breakdowns ▪ Move forward aiming at higher level of production technology	▪ Firmly establish self-supervision to be able to develop factory's strategy by PM groups themselves without managers' detailed instruction ▪ Detect and resolve arising problems by PM groups themselves by way of short remedial program	▪ Assist activities to maintain, improve and hand down current TPM status ▪ Encourage further improvement of technical knowledge and skills ▪ Move forward toward the second generation of TPM

Establish basic equipment conditions. By concentrating all efforts on cleaning and lubrication, the basic equipment conditions are established. As a result, an adequate system can be organized to maintain concurrent operating conditions. From an equipment perspective, it aims at restoring deteriorated parts and preventing forced deterioration.

Step 1: Initial cleaning

- Thoroughly clean equipment and its surroundings. Remove any foreign substances and unnecessary materials that negatively influence quality and equipment.
- Recognize the harmful influence of contaminants in terms of safety, quality, and equipment.
- Detect and remedy any deteriorated and defective areas in equipment.
- List the sources of contamination and difficult cleaning areas.

Step 2: Countermeasures to sources of contamination

- Review the sources of contamination and difficult cleaning areas listed in Step 1 in terms of the factors that influence safety, quality, and equipment.
- Set tentative cleaning standards.
- Take remedial actions in order to complete cleaning within a targeted time.
- Learn about safety and quality, and principles of processing by taking remedial actions against sources of contamination.
- Alter difficult cleaning areas to be able to finish cleaning within a targeted time.

Step 3: Cleaning/lubricating standards

- Conduct lubrication education for operators.
- Check all lubricating points and surfaces.
- Set tentative lubricating standards.
- Detect and remedy deteriorated and defective parts in equipment in terms of lubrication.
- Remedy difficult lubricating areas to be able to finish lubricating within a targeted time.
- By combining the tentative cleaning standards set in Step 2 and the tentative lubricating standards set in Step 3, tentatively set cleaning/lubricating standards.
- Clean and lubricate equipment as a whole in accordance with cleaning/lubricating standards.

- Improve work methods and equipment to be able to finish cleaning/lubricating within a targeted time.
- Revise cleaning/lubricating standards in response to the results of remedial actions.

Understand what autonomous supervision is. The activities just mentioned can never be satisfactorily achieved on the shopfloor unless all employees are involved. Throughout these three steps, in the first stage of the program, all personnel involved learn about the CAPD cycle by practice. Operators then can understand that rules must be set by those who must follow the rules. That is, in effect, autonomous supervision.

4.2.3 The second stage of the development program (Steps 4 and 5)

The second stage of the development program consists of two steps: Steps 4 and 5. The major activities to be carried out here are an overall category-by-category inspection and the establishment of operators' routine maintenance system. This stage focuses on achieving a dramatic reduction of breakdowns and minor stoppages, along with training knowledgeable operators through the repetition of education and subsequent practice of inspection.

The activities programmed in the second stage are the various remedial actions used to inspect for and restore all minute deteriorated and defective parts in equipment so as to attain Zero Breakdowns. Figure 4.2 is an example of how Zero Breakdowns are attained in an automobile manufacturing plant. The major objectives are to "conduct overall inspection of equipment defects" and "train knowledgeable operators."

Step 4: Overall inspection

- Repeat the activities listed here in accordance with overall inspection categories such as fastener, electrical, power transmission, hydraulics/pneumatics, and so on.
- Conduct roll-out education for operators by using inspection check sheets, manuals, and other teaching materials prepared for each category.
- Inspect and remedy deteriorated and defective parts.
- Identify difficult inspecting areas.
- Set tentative inspecting standards.
- Restore deteriorated and defective parts in equipment.
- Modify difficult inspecting areas.

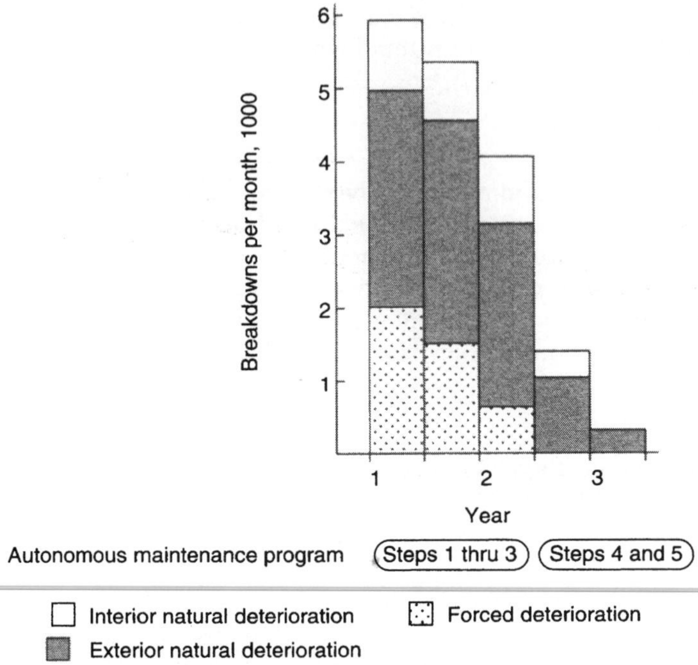

Summary of plant

Production capacity: 35,000 cars per month
Number of employees: 3,200
Number of equipment: 4,000

Figure 4.2 The challenge of Zero Breakdowns.

- Alter inspection methods and equipment by applying visual control to finish inspecting task within a targeted time.
- Revise tentative inspecting standards.
- Repeat the CAPD cycle during a short interval so as to master autonomous supervision.

Step 5: Autonomous maintenance standards

- Compare the inspection standards set by autonomous maintenance in Step 4 with those standards which are prepared by full-time maintenance in order to allocate routine inspecting tasks between the production and maintenance departments.
- By combining the cleaning/lubrication standards set in Step 3 and the inspecting standards set in Step 4, set tentative autonomous maintenance standards and schedules.
- Clean, lubricate, and inspect equipment as a whole according to tentative autonomous maintenance standards and schedules.

- Alter work methods and equipment in order to finish routine cleaning, lubricating, and inspecting task within a targeted time.
- Apply visual controls more thoroughly and revise autonomous maintenance standards so as to become more effective.
- Finalize autonomous maintenance standards in response to the results of remedial actions.

4.2.4 The third stage of the development program (Step 6)

The third stage of the development program is Step 6. In the previous steps, autonomous maintenance focused on equipment to attain Zero Breakdowns. In this stage, operators' efforts to achieve process quality assurance promote the attaining of Zero Defects.

Step 6-1: Remedies focused on quality results

- Identify process quality in connection with each process or piece of equipment. Draw quality assurance flow diagrams.
- Review process quality and evaluation criteria based on the five criteria for ease of observation.
- Improve equipment and work methods thoroughly by applying visual controls and mistake proofs until the five criteria for ease of observation are met.
- Classify the outflow of defective products in terms of customer satisfaction, frequency of occurrence, possibility of outflow, damage and confusion in the downstream processes, and so on.
- Review quality and defective product handling.
- Check whether the occurrence of defective product can be detected at a glance.

Step 6-2: Remedies focused on quality causes

- Identify the quality conditions for each process or piece of equipment.
- Review the quality conditions based on the five criteria for quality assurance.
- Improve equipment and work methods until these same five criteria can be met.
- Review process quality, quality conditions, inspecting standards, and so on.

Step 6-3: Establish process quality assurance system

- By making a thorough assessment of process quality and quality conditions, operators maintain processes in which either the occurrence of quality defects or the outflow of defective products to downstream processes are definitely prevented.

4.2.5 The completion stage of the development program (Step 7)

The autonomous maintenance program is finished in Step 7. The CAPD cycle is firmly implemented on the shopfloor whereby operators can develop corporate policy and factory targets by themselves, or, in other words, by way of autonomous supervision. Operators continuously realize and maintain optimal plant conditions by setting the rules that they must follow.

4.3 The Step-by-Step Development of Autonomous Maintenance

4.3.1 The CAPD cycle

All autonomous maintenance programs are structured so as to repeat the CAPD cycle. Although the same words used in Deming's Cycle—Plan (P), Do (D), Check (C), and Act (A)—are used in order to facilitate understanding, the CAPD cycle in TPM involves quite different concepts, as illustrated in Fig. 4.3 and described here.

- Check (C): Thoroughly examine the status quo and expose the problems.
- Act (A): Take countermeasures to solve the problems.
- Plan (P): Prevent recurrence of problems by improving equipment in so far as it may be cost effective to do so. If efforts fail, invent visual controls to detect problems at a glance. If both remedial actions fail, prevent the recurrence by human intervention. Prepare the rules to be followed, such as work standards, procedures, and check sheets.
- Do (D): Execute and follow the rules to prevent the recurrence of the same problem, and return to Check (C) if previous efforts are not sufficient. Repeat the same CAPD cycle until the targeted criteria is satisfied.

In the autonomous maintenance program, operators repeat the CAPD cycle many times. Therefore, they master autonomous supervision. Figure 4.4 shows the repetition of the CAPD cycle focused on equipment to attain Zero Breakdowns.

This concept is very effective not only in equipment-oriented production departments, but also in manual work-oriented departments, such as assembly, quality assurance, and material handling, or in other administrative departments. All TPM programs are constructed on the basis of this same concept. It is especially essential to be familiar with the CAPD cycle in order to understand autonomous maintenance and to plan one's own TPM development program in any department.

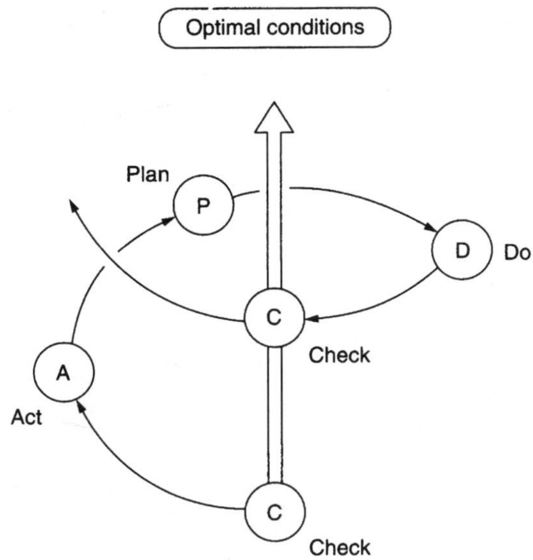

Figure 4.3 The repetition of the CAPD cycle aiming at the realization of optimal conditions.

4.3.2 Managers must take the lead and demonstrate by example

Prior to the initiation of each step by PM groups, the manager's group must execute the scheduled stepwise program with the manager's model. Using actual equipment, frontline managers and engineers come to understand by trial and error how to proceed in which area of equipment and to what extent. Without these experiences involving their own equipment, they cannot understand how to supervise and train operators in accordance with the TPM concept, and finally how to conduct the autonomous maintenance audit. On the other hand, PM groups, by observing managers' activities during routine operations, can roughly anticipate the programmed activities in the forthcoming step.

What managers are essentially expected to do is to provide a good and clear example with a given piece of equipment. No matter how excellent written or audiovisual materials may be, they cannot be a substitute for physical examples in demonstrating how cleanliness is targeted, for instance, in the initial cleaning per Step 1.

Meanwhile, once the autonomous maintenance program is launched, the heaviest burden falls on frontline managers, sometimes to the point of overload. As a result, in some cases, they are not able to maintain their position as role models. At the discretion of the manager's model, equipment is carefully selected so as to provide a feasible level of complexity as a demonstrative pilot model. The ideal piece of equip-

Step		CAPD cycle	Major activities
1. Initial cleaning		(C) → (A) → (P) → (D) → (C)	• Thoroughly clean equipment. • Locate and remedy defective areas in equipment. • Identify sources of contamination.
2. Countermeasures to sources of contamination			• Take countermeasures against sources of contamination and difficult cleaning areas.
3. Cleaning and lubricating standards			• Conduct overall inspection for lubrication. • Set cleaning/lubricating standards.
4. Overall inspection	4-1	(C) → (A) → (P) → (D)	• Conduct overall inspection for fasteners. • Set tentative inspecting standards.
			• Correct minute defective areas in equipment. • Take countermeasures against difficult inspecting areas.
			• Revise and follow routine inspecting standards. • Improve further equipment and inspecting methods.
	4-2	ditto	• Repeat procedures as above for electrical components.
	4-3	ditto	• Repeat procedures as above for power transmission.
	4-4	ditto	• Repeat procedures as above for hydraulics/pneumatics.
5. Autonomous maintenance standards		(C) → (A) → (P) → (D) → (C)	• Totally clean, lubricate, and inspect equipment.
			• Identify difficult work areas and take necessary remedial actions.
			• Set autonomous maintenance standards and schedules in terms of routine cleaning, lubricating, and inspection.
			• Follow autonomous maintenance standards and schedule. • Improve further equipment and work methods.

Figure 4.4 Repeating the CAPD cycle focused on equipment.

ment should not be overly complicated, but should be able to present sufficient problems for instructive purposes.

What can change the conventional way of thinking is neither theories nor logic. Long and empty discussions without thorough practice result in today's huge losses and lower productivity, despite attempts to implement many well-known concepts in the past.

It is not easy, however, to change the minds of managers and engineers who are eager to discuss strategies based only on theory and logic without a physical trial. Unless they share common experiences with operators and are familiar with actual shopfloor conditions, they can never understand how unreasonable or unfeasible the rules and requests that they impose on the work place are. During thorough practice with actual equipment and its operating conditions, they can gain valuable insight. In this way, the TPM concept can be tailored to fit the unique conditions found in each plant.

4.3.3 The autonomous maintenance audit

At the beginning of each step, a PM group prepares its action plans and schedule for the stepwise activities in accordance with observations on what and how managers did with the pilot model. Operators then develop their planned small group activities.

When this step is nearly finished, the operators compare their equipment conditions with the manager's model and conduct a self-audit. If the results of this comparison are satisfactory, the operators send a formal audit application to the TPM office. If the results of the stepwise audit are successful, they proceed to the next step, as shown in Fig. 4.5. Otherwise, they must repeat the current step. Figure 4.6 illustrates the procedure for the autonomous maintenance audit.

The autonomous maintenance audit is carried out for the sake of the final assessment of stepwise activities at the end of each step or substep. The auditors consist of frontline managers from the production department and also frequently include maintenance personnel and plant engineers. Many visitors from other PM groups and departments may participate in the audit for their own better understanding of audit procedures, performance, and equipment conditions.

This audit is not an examination to select some excellent PM groups or individuals, but rather is the principal means for providing operator education and motivation. The auditors must keep in mind the following points:

Prior to the audit

- In the stepwise education, teach operators how to proceed with programmed activities at specific locations of equipment and to specific extents by using actual materials and examples.

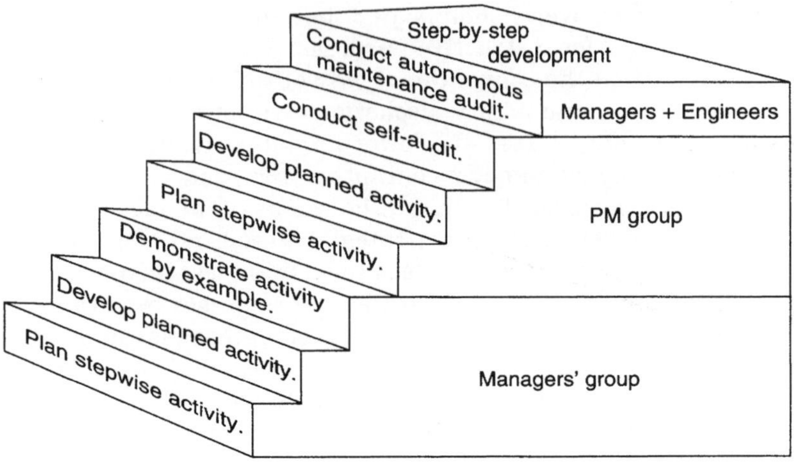

Figure 4.5 The stepwise procedure in the autonomous maintenance program.

- Clearly demonstrate the keypoints and criteria of an audit involving actual equipment in managers' models.
- Divide managers into two groups: auditors and auditees, and conduct rehearsals.
- Review audit criteria and assure even evaluation levels.

In the audit

- Indicate distinctively good or bad points in the operators' performance.
- Help operators to understand clearly how to proceed with autonomous maintenance in connection with the TPM concept.
- Review auditors' own routine supervision and advice for operators based on troubles and problems revealed in the process of the audit.
- Encourage operators by praising good results and progress.
- Make achieving a satisfactory audit a gratifying experience for operators.
- Request all operators to make some presentation.
- Ask questions of all operators and oblige each of them to answer.
- Train operators to pose questions on the spot if they do not understand something.
- Require all operators to demonstrate actual tasks in accordance with standards set by themselves.

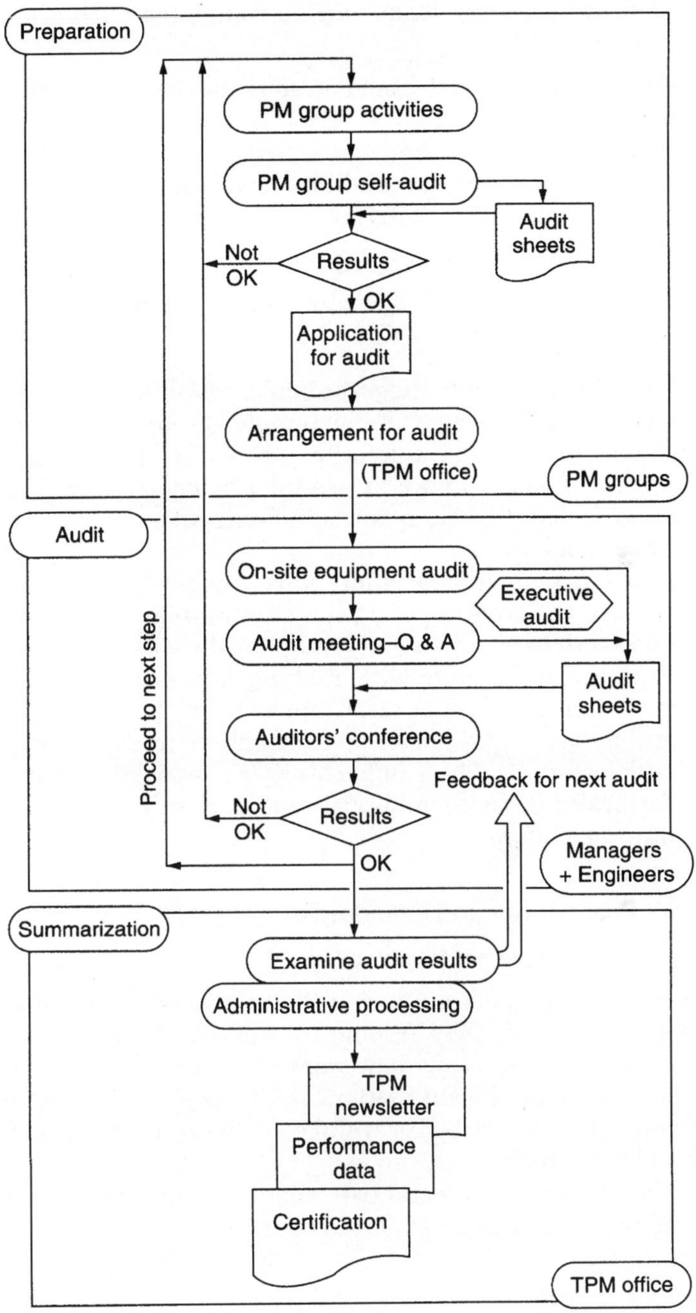

Figure 4.6 The procedure for the autonomous maintenance audit.

- Confirm that necessary tasks can be completed within a targeted time.
- Evaluate the audit results based on consultation with auditors.

After the audit

- Make an assessment of the auditors' own routine supervision and offer advice for operators to arrive at something better.
- Consider how to assist each PM group.
- Consider how to train each operator to improve technical skills (see Table 3.1).

Production department managers should not blame operators for the poor progress of PM groups. For example, group activities are not active enough, the aims of a step are not sufficiently understood by operators, the results of an audit are unsatisfactory, and so on. Such poor progress is actually the result of a manager's inadequate routine supervision and advice.

Important outcomes of the audit must include not only technical advice, but also the creation of a relaxed atmosphere for autonomous maintenance activities. At an early stage in the TPM program, almost all operators are unfamiliar with making a presentation in front of managers or other people. The audit can be helpful in getting operators to relax and in creating a congenial atmosphere in small group activities. It is, therefore, the best opportunity for educating operators, as well as for training those auditors who are anticipated to become future frontline leaders.

4.4 Work Procedures and Standards

4.4.1 Why are the rules not followed?

When frontline managers are asked the reason why the rules are not followed by operators, they frequently answer, "I have made every effort to make the operators follow work standards, but it is very hard. If there is a good idea, I want to know it." They always pressure operators to follow the rules without careful consideration of the reasons and background for doing so.

These managers must establish the following conditions before requesting their operators to follow something.

Feasibility: Confirm on their own that a given rule is feasible to follow, and demonstrate clearly how to comply with it.

Motivation: Help operators understand well why the rules must be followed, and what will happen if the rules are not followed.

Skill: Promote operator's skill.

Circumstances: Provide circumstances conducive to following directives, for instance, time, tools, or budget.

The manager's role is not one of criticizing operators all the time for failing to follow the rules. If the four conditions listed above are adequately provided, the rules will be spontaneously followed on the shopfloor. The main reason why rules are not followed is that those who set the rules are different from those who must follow the rules. As a result, operators are forced to follow or pretend to follow rules without sufficient understanding about the necessity for having them or the procedure for complying with them.

Managers or engineers should, therefore, have a try at any work standards that they set. Everyone is averse to excessive armchair theories which are impossible to follow. These assertions are true not only for the factory, but also for any workplace in the company.

4.4.2 Operators set the rules

Rules must be set by those who must follow the rules. This is the simplest and best way to assure that operators will follow them. If operators are well-trained, they can establish almost all routine work standards and procedures such as cleaning, lubricating, setup, and adjustment, except for a few that require professional technical expertise.

Convinced of the necessity for rules, operators learn how to formulate them, and then set up the rules by applying the experience they have acquired in stepwise activities in order to maintain achieved operating conditions. In this way, feasible work standards can be prepared by operators themselves, and clear definitions of the "5W's and 1H" (i.e., who, what, when, where, why, and how) can be forthcoming.

It is definitely time consuming for inexperienced operators to prepare standards by merely learning the names of parts and drawing equipment sketches. In the course of these activities, however, they come to understand the structure of equipment in more detail, and these efforts provide deep satisfaction upon completion. Rules established in this way will be followed by operators. This approach also promotes conscious participation in plant operations and highlights the operator's role. This is the first step toward autonomous supervision.

Becoming irritated at slow progress, some managers may try to arrange for a more efficient way in concert with the maintenance engineers. This kind of attempt will result only in the empty compilation of documents and a failure to train knowledgeable operators or to implement an autonomous maintenance system on the shopfloor. A sufficiently long period of time is needed for operators to develop autonomous supervision by repeating the CAPD cycle. Although it may seem like a detour, the shortest way to succeed is by the progressive and patient execution of the seven-step program of autonomous maintenance.

4.5 Educational Systems in Autonomous Maintenance

4.5.1 Types of education

In general, operators are trained in major categories of basic education as follows:

- Introductory education
- Stepwise education
- Inspection education
- Maintenance skill training
- Routine education
- Any other required educational subjects such as safety, quality, operation, and changeover

Many companies try various ideas in keeping with their circumstances. A summary of these basic educational systems follows.

Introductory education. Introductory education involves orientation and lectures on basic TPM concepts and the knowledge needed to initiate TPM activities. It is conducted by TPM instructors for factory personnel involving the plant manager throughout the preparatory stage of TPM implementation for approximately a half year. Frontline managers, supervisors, and, hopefully, PM group leaders are assembled from the workplace into a classroom according to divisional or departmental levels.

At this time, the most important task requested of frontline managers and above is the building of a consensus at each level regarding fundamental recognition on the status quo. Some factors to consider are the actual conditions of the company, the business environment, and the resulting necessity and purpose for implementing TPM.

In addition to this basic education, group leaders and operators must be instructed in depth about safety requirements. Further details of safety matters will be discussed in Chap. 5.

Stepwise education. Prior to the commencement of every step, production department managers and maintenance engineers teach operators how to proceed in each step. References are made to the purpose and the detailed procedures of scheduled activities, safety matters, and keypoints in the autonomous maintenance audit, the position of and relationship between the previous and following steps in the seven-step program, and any other requirements. Operators must be taught on actual equipment and with examples using the manager's model. It is important to emphasize the troubles, mistakes, and safety considerations by making reference to the actual experiences acquired with managers' models.

Inspection education. Common functions built into many machines, along with mass-produced parts which are widely located in each plant, are selected and sorted out into several overall inspection categories. Inspection education is conducted categorically by applying the "roll-out" education method described in the next section. Production department managers, supervisors, and PM group leaders who are taught by maintenance engineers at the initiation of each category, then, in turn, teach the operators.

The subjects selected (that is, the overall inspection categories) differ with plant conditions. Typical examples include lubrication, fastener, electrical, power transmission, hydraulics, pneumatics, and so on. Among these categories, lubrication education is carried out in Step 3, and the other categories are conducted in Step 4. More details are described in Chap. 8.

Maintenance skill training. The maintenance department, consisting of a limited number of personnel, cannot respond to the enormous number of work orders sent from autonomous maintenance. To alleviate this situation, operators need to be taught, as far as possible, relatively easy tasks such as V-belt and filter replacement, chain shortening, and welding. This approach will reduce maintenance technicians' work load and, at the same time, enhance operators' motivation.

Care must be taken, however, to avoid overloading operators with too much education and hurried practice. Otherwise, they will fail to keep up with all of the educational input by the time they complete the inspection education in Step 4. Ideally, operators progress gradually with the assistance of a group leader and maintenance personnel.

Routine education. Even though it is the key to success in TPM, the time available for education is limited in a busy operator's routine. Instruction cannot be, however, limited to a specifically designated time. Therefore, during daily meetings or on the job, group leaders, and sometimes managers or maintenance personnel, repeat the same subject in five- or ten-minute spans until it is well understood by all operators. From the simplest to the most complex, all subjects required to produce knowledgeable operators are taught patiently and continuously. Particular emphasis must be placed on severe breakdowns and misoperations that occur in the operator's own plant.

4.5.2 Means of education

In TPM, "roll out" education and the one-point lesson are used as the major means of routine education for operators.

Roll-out education. To begin with, PM group leaders are trained by maintenance personnel or production managers. Then these group leaders transmit their newly acquired know-how to operators. Such indirect education for operators is referred to as roll-out education. In this type of education, group leaders are expected to enhance their own leadership skills by teaching operators.

Furthermore, unless leaders themselves come to thoroughly understand subject matter during the course of their own training, they will never be able to teach it to operators afterwards. Frequently, they need to seek the advice of managers and maintenance personnel, and to study on their own. These leaders must be particularly adept at teaching about the functions and characteristics of the specific equipment operated by their own PM group. This type of obligation promotes a sense of responsibility in the group leader's mind. Roll-out education is, in these ways, very effective, especially where thorough education is needed at the floor level in a large managerial structure.

The one-point lesson. Any topic can be selected as subject matter about the structure and function of equipment, removal and reinstallation of jigs, cleaning, lubrication, methods and criteria of inspection, safety, and any other matters of interest. As shown in Fig. 4.7, only one point is illustrated on a sheet of paper. At every opportunity, when all group members gather, that one point is explained over and over again. The "one-point lesson" refers to this one-subject/one-sheet training method, and is most frequently conducted at the work place.

One-point lessons are generally prepared by frontline managers, supervisors, group leaders, and sometimes by operators. Those who do prepare them are requested to have a basic understanding of and accurate knowledge about the selected topics. They must study on their own and consult with specialists. Teaching materials focused on their own equipment and written by their own superiors are very productive of understanding for not only operators but also writers.

One-point lessons prepared, on the other hand, by maintenance or plant engineers frequently are overly difficult for operators to understand because these lessons are likely to incorporate many technical terms and complicated drawings. All lessons, obviously, must be written as simply as possible.

4.6 Notes for Successful TPM Implementation

4.6.1 Autonomous maintenance is the job

Autonomous maintenance in TPM is developed as a job requested of all employees, in response to the specific missions assigned to each small group, according to the corporate managerial structure. Some managers, however, interpret "autonomous" as "voluntary" rather than as autonomous supervision because they are influenced by the traditional Japanese-style TQC concept. They, accordingly, believe that it is better for managers not to interfere in voluntary operators' activities. It is an absurd misunderstanding, needless to say. Autonomous maintenance is simply the means to attain corporate targets.

To be able to exercise autonomous supervision in the workplace, every employee must have the required motivation, skills, and circumstances. Providing the favorable circumstances is, essentially, a manager's role. From this point of view, managers must make the largest contribution to the enhancement of motivation and skills of frontline personnel, along with fulfilling their own daily duties.

4.6.2 The short remedial program

As a topic in a short remedial program, PM groups select a suitable problem from the six big losses that recur in their equipment. According to a schedule unrelated to the seven-step program, operators take remedial actions to resolve the problem in a relatively short period of time. This type of program aims at enhancing technical capability and at developing habits for solving the specific problem by a PM group itself. A program is conducted as follows:

- Clearly identify the losses and comprehend the problem.
- Make an adequate analysis of the problem and pinpoint the topic.
- Describe the occurrence of losses in detail, and examine their cause-and-effect aspects based on quantitative data.
- Enhance the capability for logically analyzing and describing the causes and effects of losses based on technical theories.
- Do not apply easy remedies based on overly simple ideas.
- Record data before and after the remedial actions and review the results and effects.
- Devise measures to prevent recurrence of the same problem.

4.6.3 The PM group prepares action plans

Assume, as an example, that the production plan for the next month is to be announced at the middle of the prior month and that the time

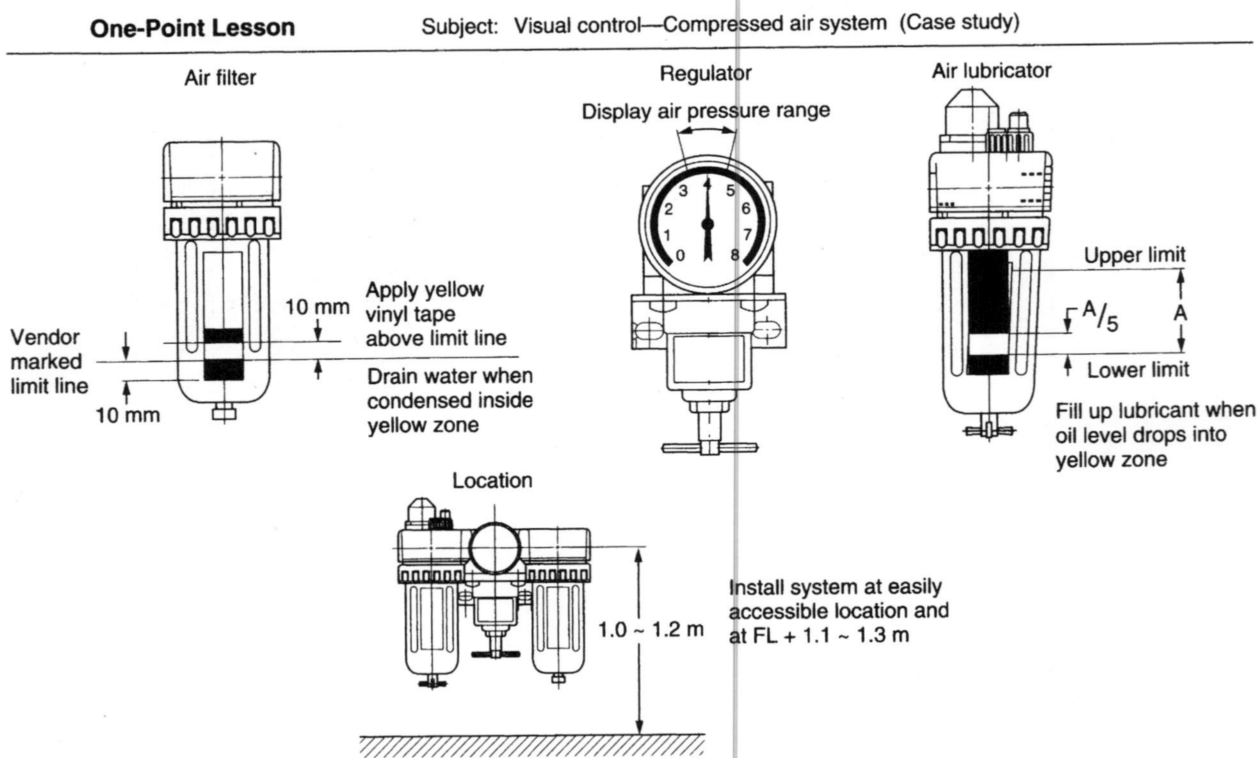

Figure 4.7 The one-point lesson.

spent for autonomous maintenance activities is targeted at 10 hours per month for each operator. PM groups must prepare an action plan and schedule for the next month in accordance with the given production plan. In some factories, the time frames for TPM activities are incorporated into the production plan in advance.

When discrepancies between the plan and actual performance occur, managers must discover the reasons:

- Insufficient time to develop the activity due to a tight production plan
- Poor planning and scheduling provided by the PM groups
- Lack of operators' enthusiasm
- Poor PM group leader's leadership

Once the background of poor progress is identified, suitable remedies must be applied as soon as possible. In this way, PM groups can practice autonomous supervision by developing group activities of their own choosing rather than those forced on them by managers. Such an approach plays an important role for operators striving to reach the status of autonomous supervision.

4.6.4 Allocating equipment by PM groups

A shift system is generally adopted in most plants. In it, the same equipment or process is operated by multiple shift crews. PM groups, accordingly, are organized for each shift. It is then necessary to clearly assign the responsibility of each group for various equipment in order to prevent confusion and duplicate activities.

Suppose, for example, that a shift consists of a crew leader and five operators who operate 17 machines in a process on a three-crew/three-shift basis. In this case, three PM groups are organized. Each group consists of five operators and is led by a crew leader. As shown in Fig. 4.8, 17 machines are divided into three A, B, and C groups, and allocated respectively to these three PM groups.

4.6.5 Encouraging autonomous maintenance activities

At least three years will be required from the launching of TPM to the time of definite implementation of Zero Breakdowns in Step 5. Another several years will be needed to actually attain Zero Defects. It is not easy to maintain the enthusiasm of all employees throughout these long periods.

Faithful execution of the seven-step program makes the largest contribution to the support of an operator's persistence. Additionally, various other ideas have been tried in many companies. Most of this book is spent on the description of these ideas. The means motivating differ with the managerial levels involved; i.e., entire company, division, or plant levels. An examination of some examples, conducted mainly by a TPM office, follows.

A TPM newsletter. A TPM newsletter is issued periodically. It addresses various topics such as case studies of improvement, accounts of efforts by operators and technicians, reports from conference participants, notices on information regarding TPM, and any other matters of common interest. The articles must, as much as possible, be written by shopfloor personnel, rather than by executives or managers. Such a newsletter is very useful for communication with, and encouragement of, employees when TPM is developed in a large scale factory, or simultaneously in several factories located at separate areas.

Conferences. In the presence of all employees, PM groups make presentations regarding successful results of autonomous maintenance, improvement activities, short remedial programs, countermeasures taken against specific misoperation, and so on. Operators can identify with and understand the progress made in other PM groups. This kind of meeting may take place once or twice a year.

An executive audit. Management personnel such as the CEO, the executives in charge of TPM, and the plant manager take plant tours and participate in an autonomous maintenance audit several times a year. The purpose is to encourage operators by showing them leadership's

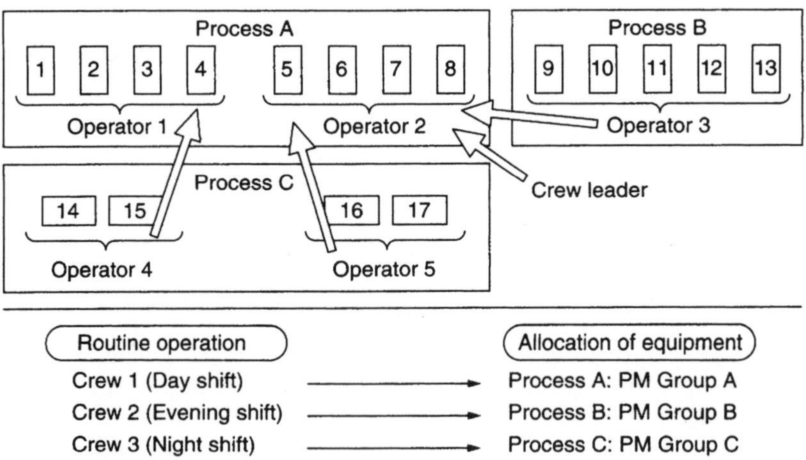

Figure 4.8 The allocation of equipment to PM groups.

deep commitment to, and high expectation of, them. These tours also ensure that leaders assess directly on the shopfloor the progress made by TPM. Leaders, during these visits, should make no negative comments in front of the operators, but should simply praise their positive results.

Events. Company-wide or factory-wide events are conducted. For example, there can be contests involving the activity board, one-point lessons, visual controls, slogans, posters, etc. These kinds of events have the effect of enhancing an employee's sense of participation in and enthusiasm for TPM activities.

The suggestion system. The suggestion system is established in many companies to make use of employees' opinions and ideas for improvement. Usually, successful suggestions resulting in outstanding improvement of preceding conditions are evaluated and awarded in accordance with the magnitude of their effects. Autonomous maintenance, however, aims at the sources of problems, and mainly at training knowledgeable operators. When an operator observes a symptom of abnormality in routine work and successfully prevents the occurrence of serious breakdowns or quality defects, such preventive actions are evaluated and awarded in much the same way as conventional suggestions. To expedite this role of operators, a warning card system is utilized in some companies to alert the maintenance department whenever any indications of unusual operating conditions are discovered. More details in this regard are illustrated in Chap. 9.

Awards and rewards. PM groups and individuals often are awarded and/or rewarded for an instance of preventing severe problems, submitting excellent improvement ideas, and winning various contests. Generally speaking, however, an honorary award rather than a monetary reward is given to encourage frontline personnel.

On the other hand, satisfactory evaluation in the stepwise audit must not be directly awarded nor rewarded. Such success must be understood as a part of the operators' routine job. Remarkable results achieved in stepwise activities may, however, be channeled through the suggestion system and awarded or rewarded.

The corporate qualification system. Operators and maintenance technicians who develop excellent skills are qualified or certified in accordance with a corporate qualification system in regard to any necessary skills in routine operation, maintenance, and safety. It is to be recommended, in this connection, to put in place definite plans for providing operator education, skill development programs, and personnel allocation.

Many other approaches may also be taken. Most importantly, each company must create its own system, in so far as possible, instead of simply copying methods developed elsewhere. Of course, no matter how concerted these superficial efforts may be, Zero Breakdowns, in the literal sense, can never be achieved.

4.6.6 Preventing dropout from TPM activities

Following the initiation of an autonomous maintenance program, gaps among PM groups in enthusiasm and progress ordinarily develop. The success achieved by groups depends largely on their group leaders' leadership and technical skills. As time goes by, the resulting gaps become larger. By watching daily group activities carefully, including information on the activity board and the operators' attitudes as reflected in the autonomous maintenance audit, managers must recognize signs of dropout as follows:

- Frequency and time spent for meetings is reduced.
- Only certain operators consistently participate in meetings.
- Excessive disparity can be found between action plans and actual performances.
- Operators are concerned with nothing but cleaning.
- Insufficient progress with remedies and preparation of standards is made because of poor technical capability.

Managers must alertly recognize these signs by addressing the following causes so as to help delinquent PM groups in various ways.

- Not enough time is available to develop the activity due to tight a production plan.
- The problem exists in a small group configuration.
- Too much equipment is allocated to the small number of operators.
- The problem exists in the group leader's leadership and technical potential.
- The problem exists in the operators.
- Because of poor technical capability, the PM group cannot understand how to proceed with improvement and how to prepare standards.

After identifying such problems, managers should take suitable actions as soon as possible by, for example, restructuring group activities and their scheduling, reducing the amount of equipment allocated,

TABLE 4.2 Twelve Keypoints of Autonomous Maintenance

Keypoints	Description
1. Introductory education	Conduct thorough education which includes orientation and lecture on TPM concepts prior to commencement of autonomous maintenance activities.
2. Cooperation among departments	Promote maximum cooperation among production-related departments as well as administrative departments. Managers must establish a support system for operators' efforts.
3. Autonomous maintenance is the job!	All employees must recognize autonomous maintenance activity as a mandatory part of operators' routine jobs.
4. Small group	All activities must be developed based on small groups.
5. Managers must take the lead!	Frontline managers must take the lead and set an example to demonstrate how to develop forthcoming steps of autonomous maintenance program.
6. Education and practice	Conduct thorough education and practice for operators without missing any minor opportunity.
7. Practice first	Take breakthrough approach by way of thorough practice in order to attain Zero Accidents, Zero Defects and Zero Breakdowns.
8. Actual effects	Provide concrete subjects and targets for operators in terms of each TPM activity, and encourage them to attain actual and effective results.
9. Rules set by operators	The rules must be set by those who must follow them.
10. Autonomous maintenance audit	The autonomous maintenance audit makes the largest contribution toward encouraging and training PM groups.
11. Quick response	The maintenance department must quickly and promptly treat work orders from autonomous maintenance. If not, PM group activity will certainly fail.
12. Be thorough	Be thorough in developing each step of autonomous maintenance program. If an audit is unsuccessful, do not proceed to the next step in a hurry because of the schedule. When this happens, TPM is not firmly implemented due to poor progress in technical knowledge and skills.

NOTE: The seven-step program repeats keypoints 5 through 10 as many times as possible.

correcting technical weakness, and reinforcing maintenance personnel's assistance. All of these considerations point out the necessity for managers, supervisors, and engineers to stay one step ahead of the managers' models. Without sufficient experience and planning, these same personnel can neither detect signs of potential problems themselves, nor can they provide effective supervision and advice for operators.

4.7 The Twelve Keypoints of Autonomous Maintenance

Prior to a detailed discussion of the autonomous maintenance program, the twelve keypoints for its successful implementation are summarized in Table 4.2. If any one of these keypoints is not adequately addressed, the dedicated efforts of frontline personnel can be expected to fail.

Chapter

5

Step 1: Initial Cleaning

5.1 Aims from the Equipment Perspective

5.1.1 Initial cleaning

Initial cleaning refers to efforts to remove completely at the commencement of autonomous maintenance activities any foreign substances such as dirt, dust, chips, grease, sludge, and scraps that adhere to equipment, dies, tools, and jigs. It is not the usual cleaning as understood in the traditional sense of factory management, but rather it entails the overall cleaning of equipment until it is thoroughly free from all kinds of contamination.

Not only managers and operators in the production department, but also all employees involved in supporting areas, such as the maintenance, plant engineering, and quality assurance departments, must learn by practice on actual equipment the meaning of exposing hidden defects by cleaning and the value of paying serious attention to even minor defects.

In the previous chapters, the importance of minor defects and cleaning is emphasized. Some damaging effects of insufficient cleaning are discussed here.

1. Foreign particles adhering to or intermingling with sliding, hydraulic, and pneumatic parts and electric or instrumental devices cause abnormal friction, vibration, erosion, clogging, leakage, burnout, or insulation deterioration. These conditions result in inaccurate machining or defective products, along with malfunctions and breakdowns of equipment.
2. Chutes contaminated with foreign materials interfere with the automatic workpiece supply to machines and thereby cause minor stoppages or quality defects.

3. Contamination frequently, directly, and negatively affects quality as noted below:
   - In the precision machining process, foreign substances adhering to the fixture of tools and jigs cause difficulties in alignment and result in quality defects due to misalignment during the run.
   - In the assembly process of electrical parts, foreign particles adhering to equipment contaminate workpieces and result in faulty contacts.
   - In the molding process of plastic products, foreign particles, adhering either to mold or pressure screws, or mixing with raw materials, cause leakage of melted resin at the mold fixture or carbonization of resin in the heating chamber.
   - In the decorating process of metal cans, foreign particles adhering to rollers assembled in inkers cause incomplete decoration or uneven color.
   - In the baking oven of an automobile paint shop, foreign particles enter from the outside, adhere to or, after becoming a condensed vapor, drop onto the automobile body and result in defective painting.
4. Contaminated equipment, tools, and jigs are not effectively inspected with the naked eye. It is almost impossible by gross observation to locate minor defects such as abrasion, looseness, scratches, deformation, leakage, and so on. Everyone eventually gives up and overlooks equipment soiled by layers of contamination and leakage. In the worst cases, nobody even gets near the equipment, let alone inspects it.

5.1.2 Cleaning is inspection

From the TPM perspective, as emphasized in Chap. 3, cleaning is aimed at exposing and eliminating hidden defects. If only a clean plant or piece of equipment is desirable, cleaning can be relegated easily to an outside subcontractor without the implementation of the pervasive measures prescribed in an operator's autonomous maintenance.

During cleaning, everyone touches all parts of the equipment and takes a look in every nook and corner. This thorough approach increases the chances of detecting hidden defects, abnormal vibration, noise, odor, and overheating. Because defects are easily detected in equipment free from contaminants, it becomes possible to remedy quickly minor and medium defects before they grow into major defects and result in breakdowns or quality defects. This is the meaning of "cleaning is inspection" as diagramed in Fig. 5.1.

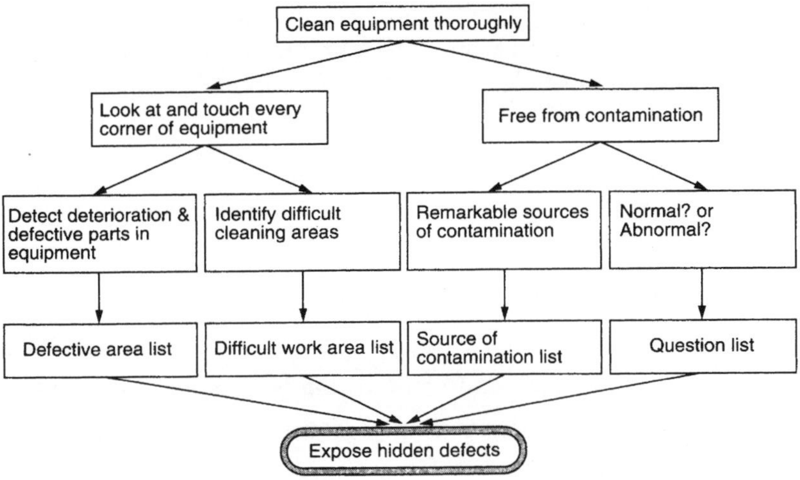

Figure 5.1 "Cleaning is inspection" and the four lists.

Cleaning is the starting point of autonomous maintenance activities. It remains the point of inception even after successful TPM implementation. Remembering at all times that "cleaning is inspection," operators clean equipment in order to prevent breakdowns and quality defects. As one of the side effects, the plant is kept clean.

5.2 Aims from the Human Perspective

No matter how well it is cleaned, equipment becomes covered daily with dirt, dust, grease, and chips. The same areas must be cleaned repeatedly many times. During the initial cleaning in Step 1, such a situation may exist until remedial actions are taken in Step 2.

Many people become dissatisfied with the course of events and complain, "We are not working in our plant to clean like a janitor," "I am sick and tired of cleaning the same place everyday," "Cleaning the inside of equipment is a maintenance technician's job, isn't it?," etc. Similar conversations arise among operators as well as managers and supervisors, who must be taught to arrive at a basic understanding of TPM concepts by way of introductory education.

Frontline personnel, including managers, may feel angry and be inclined to grumble. Such attitudes are desirable at this stage. Sources of contamination are clearly visible and attract their attention because all other areas are fully clean. Only this experience teaches them the importance of cleaning and the necessity of remedies. It is, therefore, essential to examine critically the traditional views and commonsense approaches on the shopfloor. Because it is very difficult for anyone to change a long-standing way of thinking, a sufficiently tough experience stemming from grimy conditions is needed to produce a breakthrough point.

In order to attain Zero Breakdowns, it is primarily a matter of revising fixed ideas and behavior in traditional facility management. Managers, supervisors, and PM group leaders, those who supervise operators, must take into account the aims of autonomous maintenance from human as well as equipment perspectives, and inquire of themselves as follows:

- What must be taught to each operator?
- To which areas do operators apply acquired skills and motivation?
- How can operators' skills and motivation be further developed?
- How can frontline managers help operators do the above?

5.2.1 Establish familiarity with small group activities by way of easy tasks

In TPM, small group activities are developed at each level of the managerial organization within an environment of total employee involvement. To those who are experienced in quality circle activities following conventional QC concepts, an acceptance of this type of group activities may be relatively easy. Even so, a certain amount of time is needed to become familiar and comfortable with the TPM principle of "practice first."

In this regard, frontline managers and engineers are similar to operators. Despite how well they may have learned the TPM concepts in terms of theory and logic during introductory education, there are many aspects of TPM that are difficult to comprehend without actual practice.

Anyone, who is eager and has time, can join in the cleaning activities. In promoting familiarity with and total involvement in small group activities, a prerequisite is starting with an easier task in which all employees can participate. From various points of view, it is most feasible to start an autonomous maintenance program with cleaning.

The task of cleaning equipment which has been continuously dirty for a long time demands an enormous effort. Moreover, it requires arranging for downtime of equipment during routine production. Some factories have spent a half to one year in this thorough cleaning of the entire plant. Because the group leader alone or some of the operators on their own can never finish this kind of undertaking, all operators in the PM group must contribute to the effort.

The primary role of PM group leaders is to provide leadership and build teamwork among the operators. Group leaders must encourage operators to participate in TPM activities by consent, rather than by force. This consent ideally should be based on operators' appreciation of autonomous maintenance and its implications. Beginning with the relatively easy task of cleaning, group leaders learn to master the management of small groups of operators. During this process, operators also come to know each other as members of small groups participating in a common activity.

5.2.2 Enhance operators' motivation to take care of equipment

During initial cleaning, operators might experience, for the first time, a close examination of all areas of their equipment. Operators who are, at first, reluctant to clean, gradually develop a genuine interest in their equipment by way of repetitive cleaning, followed by group discussions. They, as a result, naturally want to keep their equipment clean. Concurrently, many discoveries and new questions come forth, such as:

- What kind of malfunctions will be caused by this dirt and dust?
- Where are the sources of these contaminants? How can they be eliminated?
- Isn't there an easier way to clean?
- Are there any loose bolts, worn parts, or other defective parts?
- How do these parts function?
- If this piece of equipment breaks down, does it require a long time to fix?
- Are present operating conditions of this piece of equipment satisfactory? Are they normal or somewhat abnormal?

All operators suggest answers and countermeasures to these questions in a group meeting. If they do not resolve the matters by themselves, they ask the advice of managers or maintenance personnel. In this way, operators spontaneously increase their interest in and motivation for taking care of their equipment. They, as a result, gradually learn how to solve problems by themselves or, in other words, approach the status of autonomous supervision.

5.3 How to Develop Step 1

5.3.1 The activity board and the "four lists"

Most plants are operated on a shift basis. Operators who are in charge of the same process or equipment on different shifts have very few opportunities to meet. In order to maintain better communication among these personnel, each PM group prepares a large display board called an "activity board," in order to communicate the present conditions of and the progress achieved by each group.

Because the board aims at operators' better understanding of each other, excessive efforts at ornamentation are not necessary. For the activity board to be effective, it must allow the viewer to quickly discover the various group activities and pick out the progress noted. Particularly helpful to other groups is the posting of excellent ideas for improvement and one-point lessons.

These documents, as displayed on the board, are the most reliable sources of information by which managers can evaluate the problems and progress of all the PM groups. In an autonomous maintenance audit, the on-site meeting is begun with the operators' presentation based on the documents displayed on the activity board.

The character of the board differs from that of an operator's routine work and of the equipment allocated to each PM group. The basic information to be provided by all groups must include the following items:

- The theme and target of the activity, action plan, and schedule set by the PM group
- Bench marks, targets, actual performances, and trends in terms of major indexes measuring equipment effectiveness and the six big losses
- Planning and actual performance of group activities (meeting hours, frequency, subject, participants, and number of suggestions)
- The four lists (defective area list, question list, source of contamination list, and difficult work area list)
- Photographs before and after major actions
- One-point lessons
- Suggestions for improvement
- Progress in short remedial programs and a summary of activities
- Safety matters
- Any other information to be noted

Here follows an elaboration on the four lists, which play the most important role among the items mentioned above, as illustrated in Fig. 5.2.

1. *Defective area list*

   For the sake of definite identification, operators must tag each defective area of equipment, such as deteriorated parts, inadequately or mistakenly assembled parts, malfunctioning compo-

Date	Defective area	Countermeasures	Taken by			Op. in charge	Sch'd date	Comp'd
			Prod.	Maint.	P. Eng.			

(1) Defective area list

Date	Question	Answer or countermeasures	Posed by	Sch'd date

(2) Question list

Figure 5.2 The four lists.

Date	Where	What	Found by

(3) Source of contamination list

Date	Difficult work area	Found by

(4) Difficult work area list

Figure 5.2 *(Continued)*

nents, or any other discrepant areas which are discovered during cleaning. Relevant information is simultaneously written onto the defective area list illustrated in Fig. 5.2 (1). Operators then decide whether to remedy the problem by themselves or to ask the maintenance department to deal with it. They also put the scheduled completion date on the list.

After the condition is corrected, the operators collect the identification tag and write the actual completion date on the list. In some plants, different colored tags are prepared according to the action taken by the production or maintenance departments. Figures 5.3 and 5.4 show a sample of an identification tag and its handling procedure. Figure 5.5 illustrates, in summary, an example of the distribution of defective areas discovered in a plant during Step 1.

As autonomous maintenance activity proceeds, a large number of work orders may be submitted to the maintenance department in connection with those defective areas which are not dealt with by operators. When any delay in response to operators' requests is anticipated, maintenance personnel must provide a clear reason for doing so. For instance, it may not be practical to shut down equipment due to a production plan or the unavailability of necessary spare parts. Excuses, such as a shortage of personnel or time, are not acceptable reasons. Because the maintenance department has a budget for dealing with such cases, it must complete the necessary work within the time frame requested by the production department, even if the excess workload is subcontracted. Otherwise, operators' morale suffers, and PM group activities lose momentum.

2. *Question list*

No matter how slight a question that has arisen during and after initial cleaning is, operators write it onto a list, as shown in Fig. 5.2 (2), called the question list. When group leaders cannot provide adequate answers to operators' questions, they seek the advice of their managers, maintenance personnel, or plant engineers.

The question list is the most useful means for operators to share questions and know-how, and to understand equipment gradually without dropping out. Questions on the list are crossed off and the completion date is recorded when all group members understand the answers by way of adequate materials, such as one-point lessons. A time limit is set for finding an answer, in order to prevent going on with regular activities without having resolved operators' problems. Figure 5.6 illustrates an example of subject matter in the questions posed during Step 1 in the same plant.

Meanwhile, everyone must be aware that some people confuse the lack of basic knowledge with the basis for asking a question. The term "question" here, in the present context, refers to an

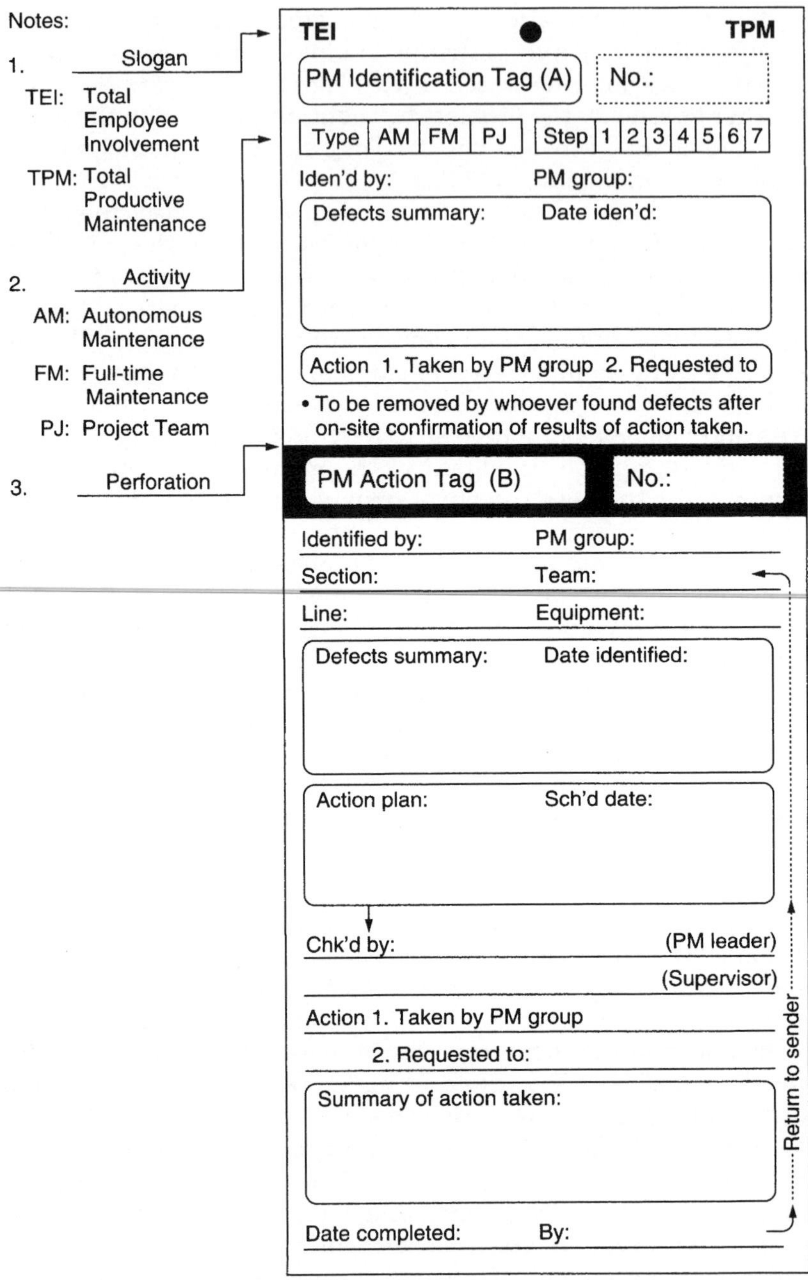

Figure 5.3 An identification tag.

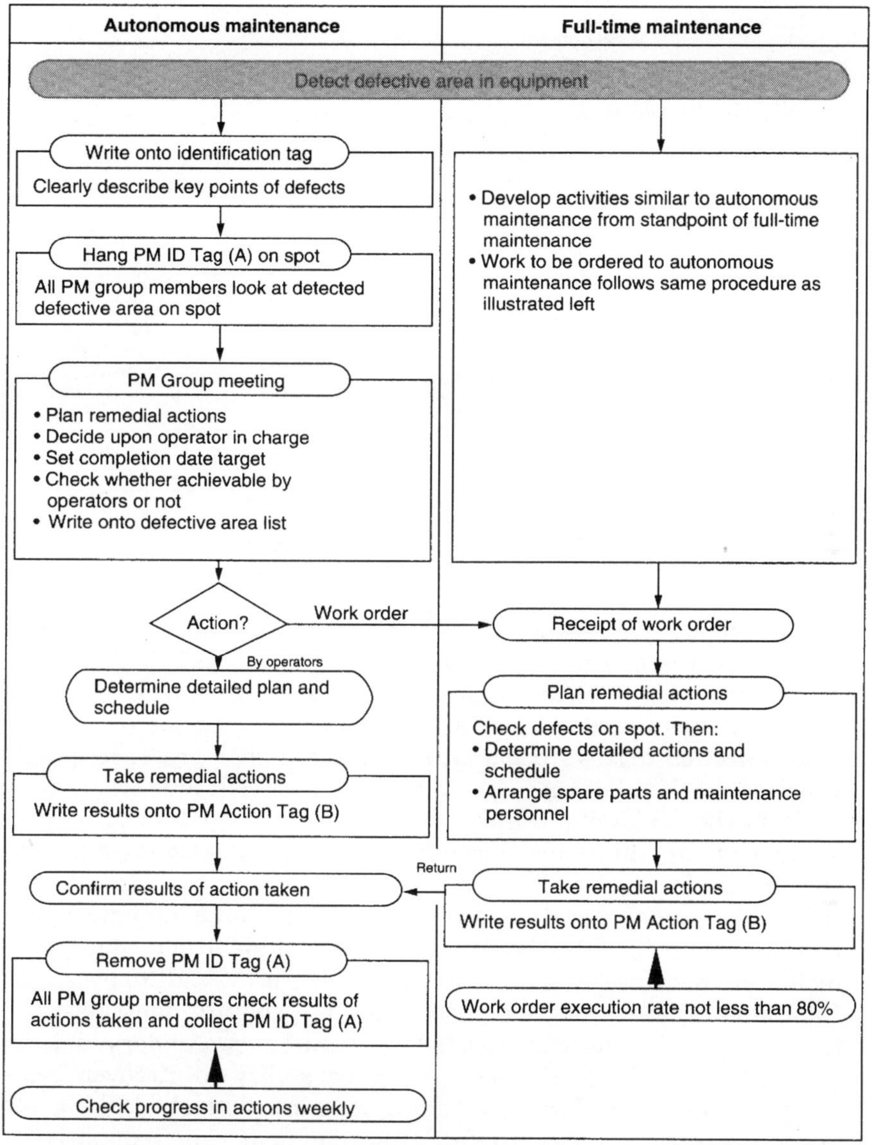

Figure 5.4 Identification tag handling procedures.

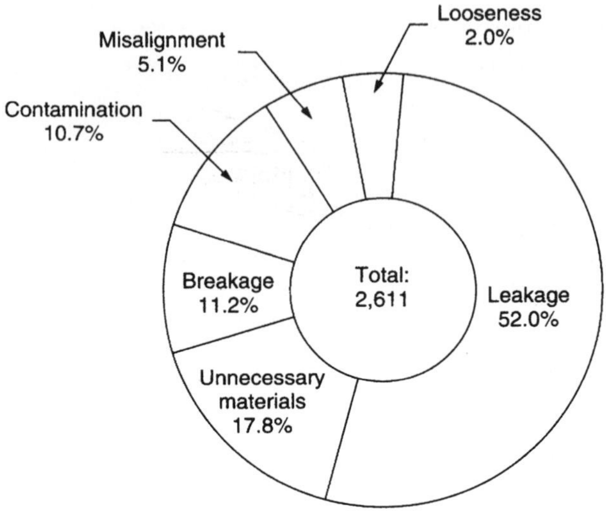

Figure 5.5 Defective areas discovered during initial cleaning.

inquiry about something unusual, which is not evaluated as defective by means of routine technical knowledge and skill. On the other hand, the inability to name a part is simply a lack of knowledge and does not constitute a "question." An example of a legitimate question arises from a situation in which a pressure gauge, which is marked between 4.5 and 5.5 kg/cm^2 with blue in order to indicate the range of normal operating pressure, points instead at 4.4 kg/cm^2. Operators are encouraged to respond with questions to such matters.

In other words, operators must develop the habit of observing and distinguishing carefully between normal and abnormal conditions, including the so called gray zone which often is difficult to evaluate. In summary, operators are encouraged to generate questions and then provide the answers that fellow operators can understand. As a result, they slowly develop a greater ability to discover more essential problems. When all employees are actively looking for questions, serious hidden defects in equipment can frequently be exposed.

3. *Sources of contamination list*
A source of contamination refers to any particular area in a piece of equipment which generates any foreign substances, such as metal chips in machining, flashes in molding, dirt and dust caused by malfunctioning equipment, leakage of lube oil and raw material, etc. Among these sources, as an instance, leakage from deteriorated parts such as a rubber hose or packing are easily remedied by sim-

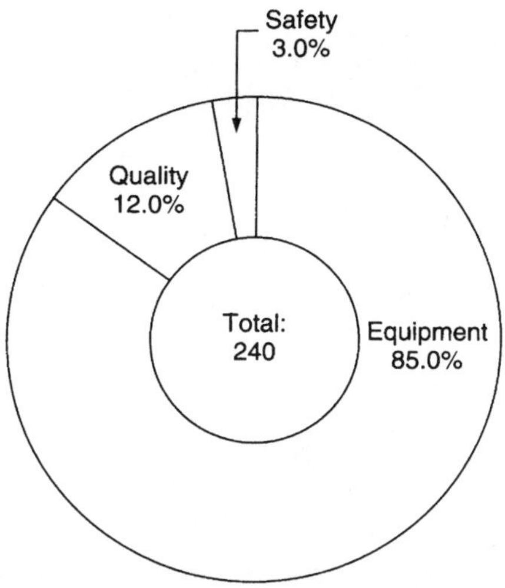

Figure 5.6 Questions posed during initial cleaning.

ply replacing or repairing the parts during Step 1. This is a typical sample of the defective areas mentioned previously.

Countermeasures to eliminate such sources, however, must be put off until Step 2 if the mechanism for generating foreign substances is complicated and needs to be studied in more detail. No easy action should be taken simply to address instantly a current difficulty with cleaning. Detected sources of contamination are recorded in a list, as shown in Fig. 5.2 (3).

In the early stage of an autonomous maintenance program, the source of contamination list is most useful in having operators understand "cleaning is inspection." Operators must make a serious and thorough examination of these sources. When sources are simply noted without further detailed observations, efficient countermeasures frequently are not taken in Step 2. Figure 5.7 details specific types of sources of contamination discovered during initial cleaning in the same plant.

4. *Difficult work area list*

A difficult work area refers to a particular area of equipment where operators experience trouble with any kind of task, such as cleaning, lubricating, inspecting, and other routine operations. These problematic areas are recorded on the difficult work area list, as shown in Fig. 5.2 (4).

Due to poor plant engineering, there are many difficult work areas which neglect the operability and maintainability of equipment, and result in rough operation as well as problems with maintenance during the later commercial production stage. These areas

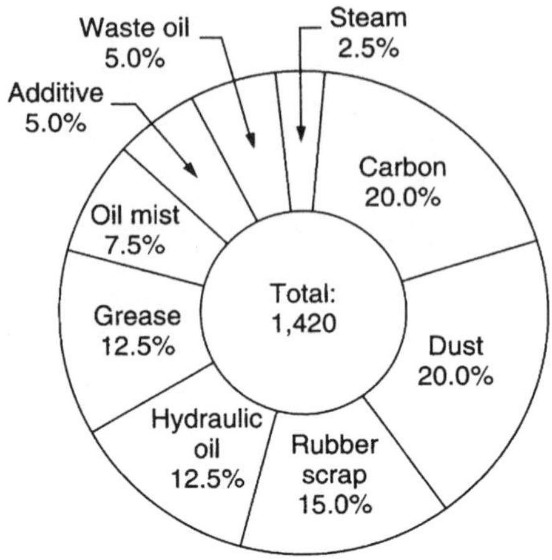

Figure 5.7 The sources of contamination discovered during initial cleaning.

usually are not recognized because almost no routine maintenance was carried out in the past.

However, when the time comes to clean, lubricate, and inspect equipment thoroughly within the limited time targets, many and various kinds of hidden difficult work areas are exposed. As the stepwise activity proceeds in keeping with the autonomous maintenance program, and is accompanied by the increasing knowledge and skill of the operators along with the press for time, even more obscure and minute difficult work areas are discovered.

When remedial actions are taken in connection with these areas, cost effectiveness always must be carefully reviewed. For instance, no actions will be required in a certain difficult cleaning area if related sources of contamination are successfully eliminated. Accordingly, countermeasures must be delayed until Step 2, except for those areas which soon become dirty again and increasingly more difficult to clean.

During the activities mentioned above, it is most important for operators to share their understanding about defective areas, questions, sources of contamination, and the difficult work areas recorded in the four lists, and to improve the group members' technical levels as simultaneously as possible. The four lists and the activity board contribute significantly to these ends.

Suppose, for example, that two hours are allocated weekly to an operator's group activity. For one of the two hours, group members actually clean equipment on site. During the other hour, they together make a detailed observation on any suitable matters of interest among those to be recorded on the four lists by way of a discussion led by the group leader. This procedure provides an opportunity to use autonomous maintenance activity as an occasion for educating the operators in their routine tasks.

5.3.2 The education accomplished by answering operators' questions is effective

In the early stages of an autonomous maintenance program, it is most important that operators not be taught overly difficult matters in a rush to achieve instant educational progress. The factory is neither a college lecture room nor a laboratory. Even though it has been quite routine for operators, as well as engineers, to operate equipment without sufficient knowledge, TPM should not be hurriedly launched in response to that situation. To the contrary, the education for operators must proceed from basic and simple matters in a stepwise fashion. Eventually, curious operators begin to learn on their own and pose questions to their managers or engineers.

The conventional approach to education by simply forcing knowledge upon students yields disappointing results with operators. No matter how primitive the operators' questions are, they must be addressed. Answering these questions is the most effective approach to education.

Operators occasionally turn to old lists of questions on the activity board and discover that they were not aware of the basic information posted there. Thus, they come to recognize their progress, which contributes to their motivation. On the other hand, frontline managers, engineers, group leaders, and maintenance personal who answer operators' questions must also study hard. This involvement in education helps raise the technical levels of these teachers and train future managers.

5.3.3 Develop countermeasures to misoperations from early on

When operators are involved in some misoperation, reprimanding them has no beneficial effects. On the contrary, most causes of misoperation reside in poor supervision and engineering rather than in operators' mistakes.

The maintenance department, therefore, must take countermeasures as early as possible to prevent the recurrence of these misoperations. In the daily meeting held between operators and maintenance personnel, such critical issues must be reviewed. If useful information of common interest results, it is compiled into one-point lessons and offered to all PM groups. In some plants, a maintenance person is assigned to each production line as an advisor.

In accordance with the assessment of misoperations that result in breakdowns, misoperations are divided into two types: the first type is caused by operator's simple carelessness and the other is caused by poor inspection, cleaning, and tightening. The latter also must be eliminated by Step 5. Figures 5.8, 5.9, and 5.10 show both a system of cooperation to prevent the recurrence of similar misoperations between the maintenance and production departments, and examples of progress achieved and remedial actions taken.

5.4 How to Proceed with Initial Cleaning

5.4.1 How to divide a step into substeps

Generally, each step is divided into appropriate substeps. Doing so helps PM groups develop the autonomous maintenance program without resorting to trial and error. The method for subdividing a step dif-

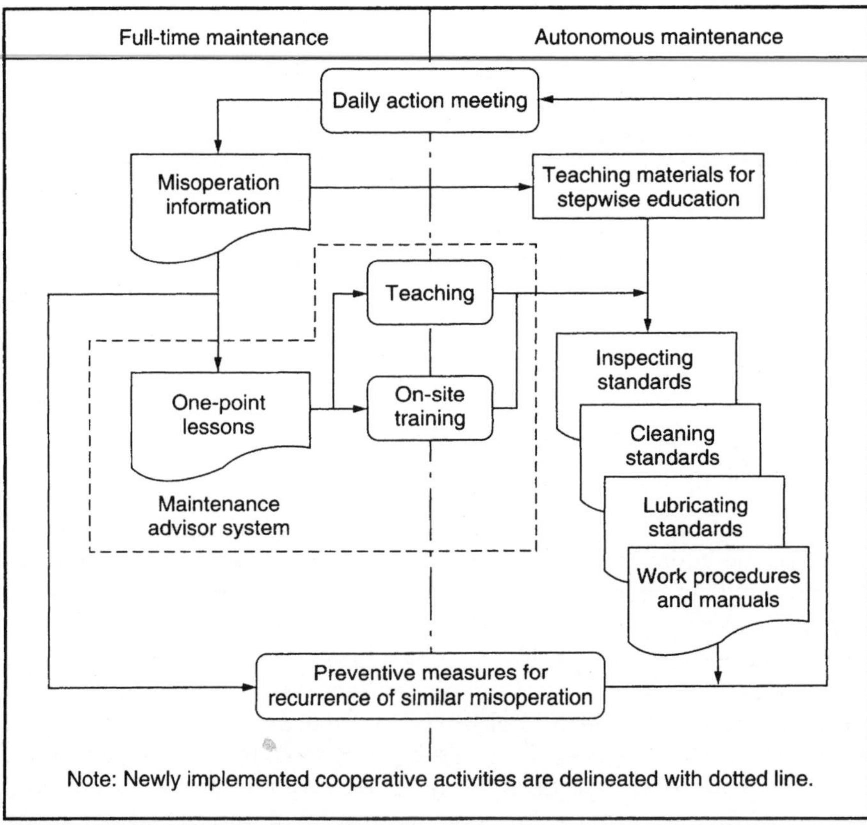

Figure 5.8 A misoperation prevention system.

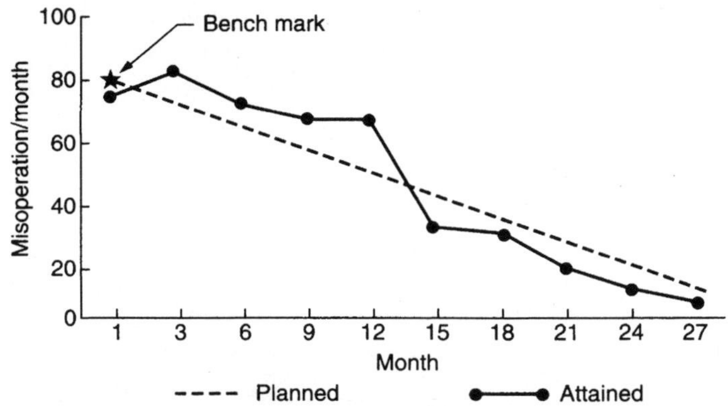

Figure 5.9 A reduction in the occurrence of misoperations.

fers from one plant or process to another, in accordance with TPM policy, configurations of processes, operating and maintaining conditions of equipment, products, operators' skills, and so on.

Especially in Step 1, some processes are found to include many sources of contamination, while others involve only a few sources, even within the same factory. Therefore, based on experience with the managers' models, managers must teach PM groups the division into substeps appropriate to the present conditions of equipment allocated to each PM group. After finishing initial cleaning, operators develop their own autonomous maintenance activities, following the substep system common to the entire production department.

5.4.2 Safety education

It might be the first experience, during initial cleaning, for most operators to lay hands on equipment from one end to the other. This same experience presents an opportunity to tie in thorough safety education for the purpose of preventing industrial accidents. Of course, this kind of introduction to safety education is not enough. Consequently, managers are required to instruct operators in detailed safety matters in accordance with actual tasks and equipment on any relevant occasion. Here is a consideration of basic safety matters:

1. *General*
   - Hard hat, dust mask, safety glasses, earplugs, and leather gloves must be worn according to the type and area of work.
   - In large machinery or vessels, operators must work in pairs under the surveillance of a watchperson.
   - Operators must not work in a dark place. Adequate lighting must be provided.

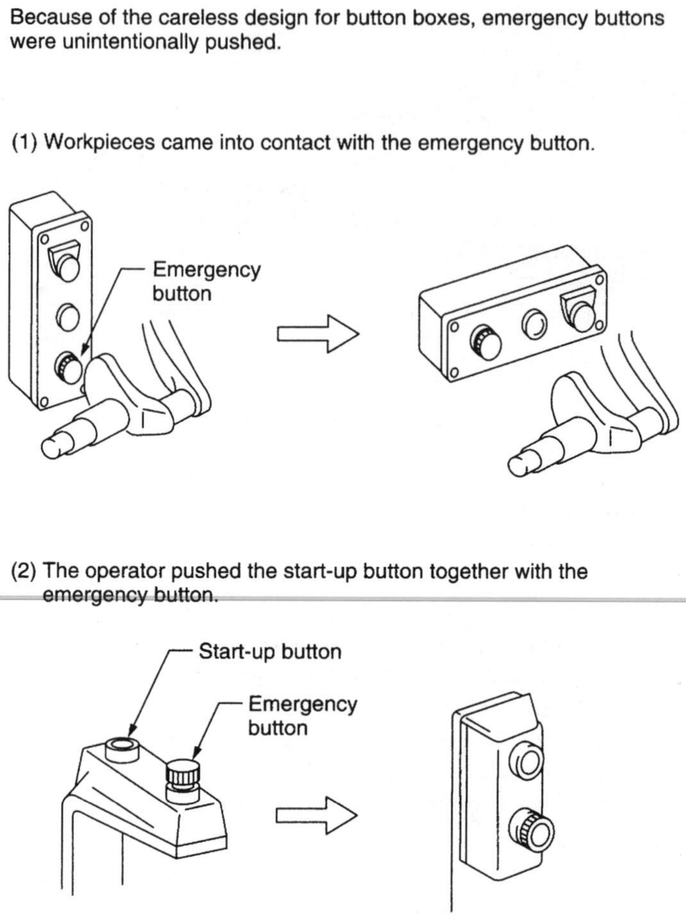

Figure 5.10 Typical remedial actions.

- Various types of instruction must be prepared and warning signs must be posted. All personnel concerned must thoroughly understand the application and significance of these aids.

2. *Work at a piece of equipment*
   - Turn off the power and put up a sign saying "autonomous maintenance at work."
   - Turn off the switch on the console and remove the key.
   - Close block valves in pneumatic or hydraulic systems, and put up a sign saying "autonomous maintenance at work."
   - Vent pressured equipment or pipes.

- At start-up and stoppage of machinery, motion to neighbors and confirm a safe status.
- Neither enter nor put hands or legs into the work area of machinery while it is operating.

3. *Work at a high place*
   - Use proper ladders or stepladders.
   - Do not stand or hang on thin pipes and cable trays.
   - Prepare scaffolds, handrails, safety nets, and lifelines.
   - Do not work above other people.
4. *Work in a vessel, furnace, or pit*
   - Confirm the oxygen level and check for flammable or noxious gas prior to and during the work.
   - Use air ventilators.
   - Post a watchperson outside the work area and put up a sign saying "autonomous maintenance at work."

5.4.3 Arrange for cleaning utensils and hand tools

Many PM groups begin similar activities simultaneously following the autonomous maintenance program. In addition, a plant-wide cleaning day or other specific activity day takes place in many plants in order to encourage frontline personnel. Cleaning utensils and hand tools may, however, frequently be in short supply because of inadequate or careless planning in the early days of TPM. If such a shortage results in operators having nothing to do and losing interest, simultaneous plant-wide cleaning activity does not make sense.

To prevent such confusion and stagnation, managers must carefully plan for and provide the required utensils and hand tools in advance. It is important to invent some special and unique tools to fit specific pieces of equipment in place of ordinary mass-produced tools.

5.4.4 How to proceed with cleaning

First of all, the most contaminated areas must be cleaned thoroughly. Then, cleaning is extended from top to bottom, center to outside, or the up to downstream sources of contamination of equipment. Special and careful cleaning must be carried out in those areas which were chronically hidden and neglected in the past; for example, equipment installed in underground pits, furnaces, ovens, ducts, and trenches.

In those facilities where airflow is especially important, such as in a paint spray cell for automobiles, a baking oven for metal coating, or a clean room for semiconductors, the unidirectional flow of interior air should be assured by design. If some experts confidently state, based on their common sense and in terms of mechanical design, that it is not necessary to pay careful attention to downstream areas, frontline personnel may easily be misled by their opinion.

Experiments using thin string or smoke, however, demonstrate that there are many swirls and much turbulence of air due to workpieces, equipment, and persons standing in the airstream. This kind of demonstration proves that the conventional approach to plant design, guaranteeing laminar flow, is a grossly inadequate. By going beyond the conventional wisdom of design and facility management, personnel expose serious hidden defects by taking a total view of conditions in the workplace and, sometimes, they achieve unexpectedly successful effects on quality.

Generally speaking, several serious defects in equipment are discovered during Step 1, as described in the case studies in this chapter. These examples must be taught in detail to all employees, so as not to miss an outstanding opportunity for valuable instruction. Motivation to clean equipment is encouraged by case studies found in one's own factory. And, finally, the importance of cleaning, especially in the context of "cleaning is inspection," is clearly recognized by all employees.

5.4.5 Thoroughly remove unnecessary components from equipment

During initial cleaning, operators discover that numerous and various kinds of obsolete components of equipment, which were removed during prior modifications, were not taken away and still remain at their original or nearby locations. Examples of remaining components are instruments, chutes, wiring, cable trays, pipes, steel structures, and, sometimes, rotary machinery and vessels. When these modifications were made, the removal of leftover components was probably neglected because of simple inconvenience, a shortage in workforce and budget, or a tight schedule. A lack of related drawings and documents, along with changes in personnel in charge, also leaves the courses and purposes of modifications unknown.

These kinds of unnecessary components of equipment often cannot be distinguished at a glance from necessary equipment in use. Someone may conclude that the cleaning of unnecessary pieces of equipment can be neglected, or that the repair of a broken glass of an unused pressure gauge is not required. This kind of attitude in facility management increases the frontline personnel's unawareness of the differences between managing facilities under operation and dealing with unnecessary components of equipment.

Sometimes, as a consequence, very dangerous situations develop, such as instruments that are still alive or pipes that are pressured, even though they should be dead or disconnected. In a plant, for example, when an operator touched an unused, but live, limit switch covered with insulating vinyl tape, a motor suddenly started and almost resulted in a severe injury.

When there is even a slight doubt about the functioning status of any item, operators should never handle it, but should request that the maintenance department deal with it.

In a chemical plant that was modified many times during its long history, the thorough removal of unnecessary parts and pieces of equipment, along with the simplifying of process configurations, resulted in remarkable maintenance cost reductions. All such unnecessary pieces of equipment should ideally be discarded completely by the end of Step 1.

5.5 Case Study

5.5.1 Typical activities during initial cleaning

In a plant where 250 operators out of 600 employees are operating 753 conventional and 97 numerical control machines, initial cleaning was conducted. Operators discovered many defective areas in equipment, sources of contaminations, and difficult work areas, as detailed in Figs. 5.11 and 5.12. During the course of these activities, 207 one-point lessons were prepared in answer to the 207 questions posed by operators. These questions and answers became an essential part of the routine educational process.

5.5.2 Severe defects detected during initial cleaning

Banbury mill is a plant preparing rubber with banbury mixers to supply all raw materials to other neighboring plants. If major equipment were to break down suddenly, production in all the other plants, in the worst case, also might shut down.

Nevertheless, the basic equipment conditions were neglected for a long time, until the entire plant was completely contaminated with a large amount of dirt, dust, rubber scraps, and greasy sludges. Neither operators nor maintenance personnel wanted to go into the underground

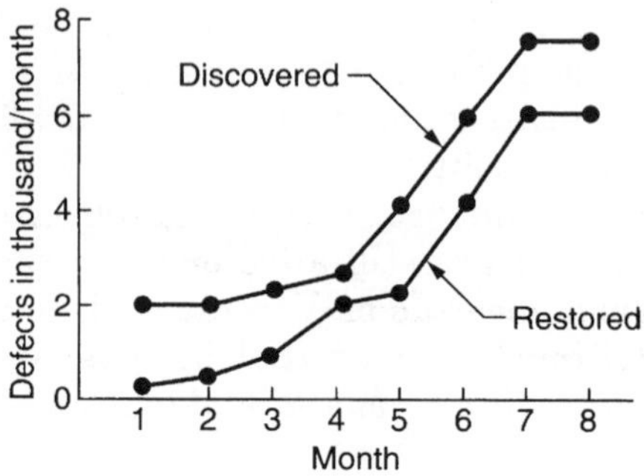

Figure 5.11 Discovering and restoring equipment defects.

pit of the banbury mixers due to the terrible contamination, even though many machines were installed there. In the event of machine breakdowns, repair work had to be preceded by cleaning of the surroundings.

The PM group in charge undertook the cleaning with great resolve. The results of this cleaning were astonishing: 2000 L of waste oil, 600 L of sludge, 200 kg of rubber scraps, 15 hand tools, 72 workpieces, and more. During this cleaning, a crack was detected in the pedestal of the Number 1 Banbury Mixer. If the crack was overlooked, the pedestal would have broken and it would have taken approximately four months to replace it. A severe effect on production was thereby prevented. There surely are many other such cases to be discovered in Step 1, as listed in Table 5.1. These cases are good examples of the expression "cleaning is inspection."

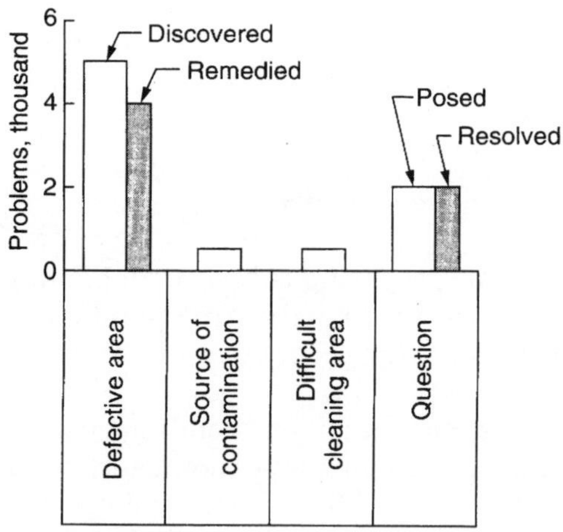

Figure 5.12 The problems written onto four lists during initial cleaning.

5.5.3 The challenge to achieve Zero foreign particles in the semiconductor industry

In a thermal diffusion process in a semiconductor manufacturing plant, cleanliness and the time needed to keep the process clean are both important. Operators must continually fight dust. Even in the high tech industries, operators' work begins and ends with cleaning. Because equipment itself involves a lot of sources of contamination, exclusive reliance on a clean room results in lower yield due to quality problems. There are many unexpected pitfalls in this regard.

Although the existence of foreign particles with diameters at the micron level, similar to particles of smoke, causes disastrous quality defects, operators were accustomed to cleaning carefully only workpiece paths without paying attention to lower areas. They were totally influenced by the fixed idea that a clean room was designed to provide downward laminar airflow from ceiling to floor.

Abandoning this past notion, they cleaned every nook and corner of their equipment, including areas enclosed by safety covers and underground ducts. Thanks to these efforts, foreign particles on silicon wafers were reduced to Zero in the coater and developer, as illustrated in Fig. 5.13. As a result, the time spent cleaning equipment was 15 percent of the bench mark, as dotted in Fig. 5.14. Furthermore, even rubber casters on chairs are considered as a source of contamination in Step 2, and appropriate countermeasures are taken. The old fixed ideas were eliminated and replaced with a new kind of awareness.

In various high tech industries, not only semiconductors, but also compact discs, photofilms, precision electrical devices, and various processes are installed in a clean room where the existence of foreign particles absolutely determines the quality and yield of products. It is, therefore, advisable when cleaning deals with particles at the micron level to question those commonsense or fixed ideas suggested by ordinary experience. This case is, then, a good example of how remarkable effects are realized by simple, but thorough, cleaning.

5.6 The Keypoints of an Autonomous Maintenance Audit

The Step 1 audit is the first and most important audit in the seven-step program and deeply influences the future development of all autonomous maintenance activities. Neither the managers who conduct the audit nor the operators who are being audited are familiar with this initial procedure. Consequently, most operators feel extremely uncomfortable. In the on-site audit and the meeting that follows, auditors must present serious questions in a humorous vein

TABLE 5.1 Serious Equipment Defects Discovered during Initial Cleaning

	No. 6 Interim mixer	No. 1 Banbury mixer	No. 6 Extruder
Location	Attachment bolts for latch cylinder	Pedestal for outlet door	Pusher casing
Defects	Bolt breakages	Minor crack in pedestal	Crack in casing
Estimated losses due to delayed discovery	Replacement of screw liner Machine downtime: 3 days	Fabrication and replacement of pedestal Machine downtime: 4 months	Fabrication and replacement of casing Machine downtime: 1 month
Corrective action taken	Bolts were replaced	Pedestal made of cast iron was reinforced by applying special staples	After temporary repairs, a new casing was fabricated and installed

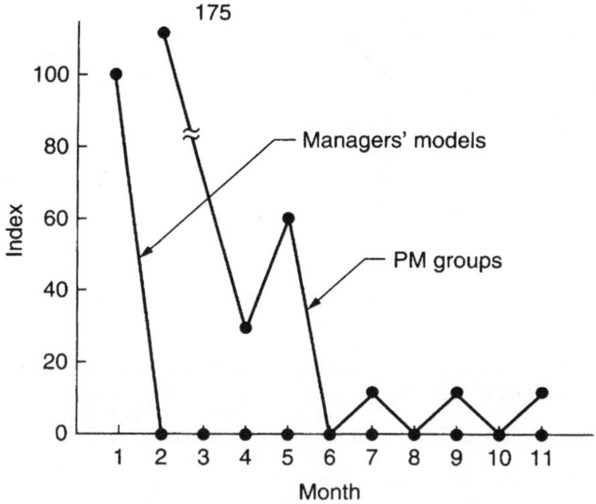

Figure 5.13 The reduction in foreign particles on silicon wafers.

to create a relaxed atmosphere. It is essential to help especially those operators who have no experience with presentations to express their opinions clearly.

Prior to the audit, auditors must thoroughly practice the procedure by trial, using the manager's model as a guide. Managers also must remember that the audit is only the final assessment in confirming of the completion of a step as well as an important opportunity for educating operators.

What is really being assessed is an auditor's leadership and managerial skills rather than an operator's efforts. Nevertheless, some managers create the audit mainly in terms of points for evaluation, which resembles a school examination with a heavy emphasis on point score and mistakes. This kind of audit does not make sense. On the contrary, it actually diminishes operators' motivation.

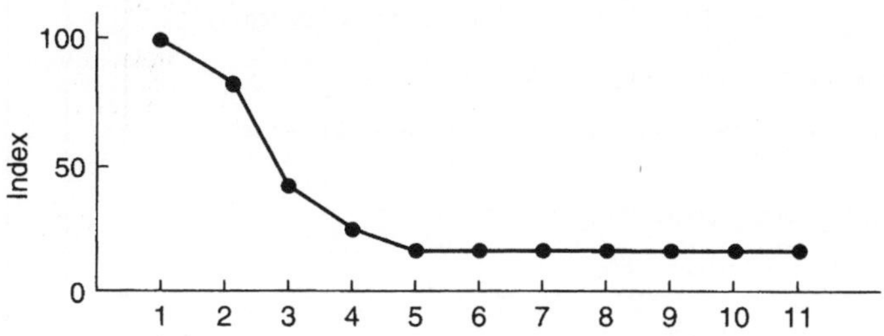

Figure 5.14 The reduction in cleaning time in vacuum processes.

TABLE 5.2

Step 1	Autonomous Maintenance Audit Sheets	Sheet 1 of 2			
No.	Audit points	Results			
1.	**Group activity** (General)				
1.1	Are aims of Step 1 understood adequately?				
1.2	Is activity plan made in advance? Well executed?				
1.3	Are managers' models well understood?				
1.4	Is activity board adequately utilized?				
1.5	Are defective areas accurately located? Confirmed by all members?				
1.6	Are previously overlooked defective areas detectable with current knowledge?				
1.7	Are identification tags collected and kept after remedies?				
1.8	Are questions sufficiently posed?				
1.9	Are questions definitely resolved? Understood by all members?				
1.10	Are sources of contamination accurately located? Written onto source of contamination list?				
1.11	Is up/downtime work clearly distinguished?				
1.12	Are safety matters carefully respected?				
1.13	Are TPM activity hours and frequency adequate?				
1.14	Is more efficient way of TPM activity pursued?				
1.15	Are used spare parts and consumables recorded?				
1.16	Is meeting after on-site activity definitely held? Reports submitted?				
1.17	Is activity participated in by all members?				
1.18	Are all members cooperating equally? Not led by particular member?				
1.19	Are noteworthy ideas introduced actively to other PM groups?				
1.20	Is cooperation with full-time maintenance satisfactory?				
2.	**Equipment** (Main body and surroundings)				
2.1	No contamination or foreign materials at locations as listed below?				
(1)	Rotating, reciprocating or positioning parts				
(2)	Parts contacting workpieces				
(3)	Frames, beds or any other member				
(4)	Conveyors, chutes or any other material transfer facilities				

TABLE 5.2 *(Continued)*

Step 1	Autonomous Maintenance Audit Sheets	Sheet 2 of 2			
(5)	Tools, jigs or any other auxiliary devices				
(6)	Sensors, microswitches, instruments, lighting (exterior/interior) of consoles/panels				
(7)	Air filters, regulators, lubricators, cylinders, solenoid valves or other pneumatic devices				
(8)	Motors, belts or safety covers and surroundings				
2.2	No deteriorated or defective parts as listed below?				
(1)	Loose, damaged or missing nuts/bolts				
(2)	Looseness in sliding parts, fixtures of tools or jigs				
(3)	Abnormal noise in motors, solenoid valves, etc.				
(4)	Damaged pipes, hoses or cables				
2.3	Are tools and jigs stored at designated locations? No shortage or damage?				
2.4	No nuts, bolts, tools, workpieces or any other unnecessary materials in and around equipment? No dropped parts on floor?				
2.5	No unnecessary pieces of equipment?				
2.6	Can quality/defective products or scraps be clearly distinguished?				
2.7	Are labels and name plates clean and legible?				
2.8	Do status display and warning lamps function properly?				
2.9	Do safety devices function properly?				
3.	**Lubrication**				
3.1	No dirt, dust or leakage in and around air lubricators, oil reservoirs or centralized systems?				
3.2	Is oil level proper? Adequate lubricants at lubricating surfaces?				
3.3	No contamination at lubricating points and surfaces?				
3.4	Are lubricants not contaminated? Not deteriorated?				
4.	**Short remedial program**				
4.1	Is outline of short remedial program known?				
5.	**Residual issues**				
5.1	No residual questions, equipment defects? If so, do reasonable explanations exist? Are plans and schedules to solve these issues clear?				

Auditors, therefore, should check whether every single operator understands group activity and is willing to participate in it. In addition, it must also be established that operators understand well how to identify defective areas of equipment, sources of contamination, and the saying, “cleaning is inspection.”

An example of an audit sheet is introduced in Table 5.2. It is necessary, of course, to prepare one’s own audit sheets to fit an individual plant and its working conditions.

Chapter

6

Step 2: Countermeasures to Sources of Contamination

6.1 Aims from the Equipment Perspective

In order to easily maintain the state of cleanliness achieved and examined in the audit during Step 1, contaminants must be eliminated at their sources. If it is absolutely impossible to remove the source, the contaminants, as a compromise, need to be prevented from dispersing by the suitable modification of equipment. When neither the removal nor the control of undesirable dispersal is at all successful, operators are obliged to clean up every source of contamination and the surroundings by hand. Ultimately, they must improve either their cleaning methods or equipment to be able to finish cleaning within time targets.

If equipment along with raw materials and products are successfully made free from contamination due to foreign substances, not only is deterioration of equipment prevented, but also the quality of products is improved. In addition, minor defects in clean equipment are easily detected and remedied before they become major defects. As a result, actions taken to prevent contamination contribute to the higher reliability of equipment. At the same time, maintainability is enhanced by means of solutions to difficult cleaning areas.

6.2 Aims from the Human Perspective

6.2.1 Learn how to solve problems and experience satisfaction with successful outcomes

In Step 1, operators clean equipment heavily contaminated through past operations over a long period of time. The extreme difficulties experienced during initial cleaning motivate operators not to allow equipment to get dirty again. Moreover, the mere existence of contaminants becomes bothersome to them once the equipment is clean. Operators and other frontline personnel, as a consequence, express a desire and an interest in making cleaning equipment easier:

- "No matter how many times I clean this machine, it becomes dirty again in a few days. I must do something to eliminate the leakage of liquid from these parts."
- "When I clean here, these chain and V-belt installations seem to be very dangerous. If the position of the motor is shifted, it should become safer. Let me consult maintenance personnel."
- "I can't afford to spend such a long time cleaning. How can I make it easier?"
- "Although we found all the defective areas in our equipment and fixed them, they're likely to deteriorate again without some countermeasures."

Step 2 promotes this kind of operators' enthusiasm. Operators actually, on their own, improve equipment based on mutual cooperation. As a result, the sources of contamination and the difficult cleaning areas are eliminated.

By means of this most familiar theme of cleaning, operators establish the basic equipment conditions and learn how to improve equipment. The resulting cleanliness of equipment contributes significantly to the elimination of breakdowns and quality defects. Thanks to the favorable outcomes obtained through the efforts of all personnel, operators comprehend the nature and significance of the TPM activities that they are developing. In this way, operators experience satisfaction when their ideas are successfully implemented.

It is Step 2 that directs operators to increase technical skill to improve equipment and enhance their confidence in proceeding to the more difficult next steps.

6.2.2 Learn about the working mechanisms of machinery

It is important in Step 2 for operators, when they prepare to take a countermeasure, to learn about the actuation and basic working mechanisms of machinery while they carefully observe the generation of foreign substances at their sources. For example:

- At a milling machine or lathe, by watching the shape, the length, and the scattering motion of metal chips, operators learn about machining metals in terms of shape, material, wear and life of cutting tools, cutting angles, feed and revolution of workpieces, and the quantity and ways of applying cutting fluid.

- At the spot welding process in a car assembly plant, operators learn about spot welding in reference to welding spots, tip shape, angle and centering of electrodes, gap of metal plates, clamping pressure, welding time, amperage of current, cooling water, and weld nugget.
- At a dust collector, operators learn about the mechanism of a cyclone to prevent dust leakage.

In observing a source of contamination that looks like a single outflow, frontline personnel, by applying a detailed where-where analysis, determine that multiple outflows concentrated into a narrow area may falsely appear as a single outflow. By investigating a phenomenon that looks like a single event through a why-why analysis, personnel discover that multiple phenomena combined with each other lead them to mistake these phenomena as a single event. The most troublesome and neglected sources of contamination have complex characteristics.

Superficial and quickly applied countermeasures, therefore, should not be used even when they are readily available. This kind of approach results in simply shifting the problem to another area. It is, therefore, recommended that a logical analysis of phenomena be conducted in view of the working mechanisms of machinery, in order to find the true solutions. Otherwise, except for instances of success by chance, foreign substances will certainly appear again at a nearby location in a slightly different way, even though the problem seems to be solved.

The aims of countermeasures against sources of contamination are not only to eliminate the generation of foreign substances and to reduce cleaning time, but also to learn about the mechanisms and dynamic motion of machinery resulting from the detailed observation of phenomena. Because no one discovered them until recently, engineers as well as operators were not able to apply effective solutions.

The approach to the source of contamination illustrates the basic concept of improvement in TPM. Moreover, the results carry over to areas of safety, quality, and equipment effectiveness. Herein lies the genuine strategy of the countermeasures against sources of contamination. In conclusion, Step 2 seeks to have everyone learn how to achieve comprehensive improvements involving familiar topics, such as outflow of contaminants.

6.3 Time Targets for Cleaning and Tentative Standards

6.3.1 Time targets for cleaning

Because operators are not always allowed to spend unlimited time for cleaning, lubricating, and inspecting equipment, managers must decide upon and announce in advance the most suitable time targets for cleaning by taking into account the configuration of the process, operating conditions, and operator allocation. For example, targets may be set at 2 minutes daily (or per shift), 5 minutes every weekend, and 15 minutes at the end of the month. Actual cleaning time, however, is almost always decided not as time required, but as time allowed for cleaning. This fact results in dirty conditions in many factories and necessitates prescribed countermeasures.

6.3.2 Prepare tentative cleaning standards and work toward improvement

At the beginning of Step 2, operators review the sources of contamination and the difficult cleaning areas discovered in Step 1. They then compile unavoidable cleaning work into tentative cleaning standards that they follow in order to maintain the cleanliness of equipment attained by the end of Step 1.

At this time, it becomes obvious which areas need to be cleaned and what the methods for cleaning are, but no countermeasures are taken yet. Operators, accordingly, clean equipment slowly if they faithfully follow the standards. Generally, they will spend much more time than the targets set by their managers. To satisfy these given targets, operators must take action against the sources of contamination and the difficult cleaning areas by improving work methods and equipment.

At the end of Step 2, tentative cleaning standards must be revised in light of the remedial actions taken for assuring the maintenance of the present cleanliness of equipment.

6.4 How to Develop Step 2

6.4.1 Substeps and the priority of countermeasures

When a source of contamination is remedied, two types of actions must take place. One is the reduction of the generation of contaminants at the source to make cleaning unnecessary. The other is the modification of methods or equipment to make cleaning easier. Of course, the latter action is unnecessary if the former is successfully achieved.

Because there are many such cases, countermeasures against sources must be conducted to avoid duplication in operators' efforts and in the cost modifying equipment, in accordance with the priorities mentioned below:

1. A remedy for the source of contamination:
   a. Remove the generation of contaminants at the source.

b. If action *a* fails, prevent the contaminants from dispersing to minimize routine cleaning tasks.

2. A remedy for difficult a cleaning area:

c. If actions *a* and *b* fail, operators clean equipment manually, and then improve cleaning methods and utensils to make cleaning procedures easier.

d. If actions *a, b,* and *c* fail, as the final alternative, modify equipment to make cleaning tasks easier.

3. Report the experience to the plant engineering and product design departments.

When the above mentioned remedies are not sufficiently effective due to current restrictions in terms of technical skill and budget, preventive measures must be taken in the future during the engineering stage of the plant and product. To this end, experience from and detailed information about remedies applied to sources of contamination should be accurately reported in a written document to the plant engineering and product design departments.

The generation of foreign substances negatively impacts safety, quality, breakdowns, minor stoppages, changeovers, and adjustments. Its countermeasures, therefore, are developed in keeping with the magnitude of these impacts, rather than with regard to the reduction of cleaning time.

Actual goals and time frames for countermeasures are decided in relation to the nature of the sources of contamination described in the following section. An example of a substep to fit specific plant conditions, illustrated in Table 6.1, should be a useful guide to planning individual strategies.

6.4.2 Types of sources of contamination

Countermeasures to sources of contamination are not restricted to minimizing the dimensions of an existing cover. Primarily, the generation of foreign substances should be eliminated at the source. If the elimination of the source is absolutely impossible to attain, the generation of foreign substances must be reduced as far as possible; their dispersion must be limited also. There are three types of sources, as described here:

1. *Generated by the technical restrictions of machinery*
   Foreign substances are definitely generated in response to the working mechanisms of machinery as follows:
   - Metal chips, scraps, and cutting fluid generated in the machining process
   - Scraps generated by a punch press

TABLE 6.1 Dividing Step 2 into Substeps

Substep	Major activity
1	Review sources of contamination.
2	Review difficult cleaning areas.
3	Prepare tentative cleaning standards.
4	Estimate cleaning intervals.
5	Set cleaning time targets.
6	Set improvement targets.
7	Sources of contamination:
7-1	Conduct a why-why analysis.
7-2	Plan remedial actions.
7-3	Take remedial actions.
7-4	Evaluate results of actions.
8	Difficult cleaning areas:
8-1	Conduct a why-why analysis.
8-2	Plan remedial actions.
8-3	Take remedial actions.
8-4	Evaluate results of actions.
9	Revise cleaning standards.
10	Assess residual issues.
11	Develop a short remedial program.
12	Conduct an autonomous maintenance audit.

- Flashes in plastic molding or rubber curing
- Solvent mist in the decorating or painting process

Against these sources, the following countermeasures must be taken:

- *Minimize* cutting allowance, quantity of cutting fluid, and evaporation of solvent.
- *Control* outflow of foreign substances from dispersing in absence of a cover.
- *Intercept* dispersing foreign substances with minimal cover.
- *Contain* foreign substances with a reservoir, gutter, or pan.

These countermeasures are very effective not only for making cleaning easier, but also for preventing various other problems. For example, reducing the cutting allowance in the cast-iron machining process is effective in lengthening cutting tool life as well as for improving yield from raw materials. These issues, however, frequently are not solved by modification of existing equipment and operating techniques, but must be approached by comprehensive countermeasures involving the plant engineering and product design departments. Similar situations are found in various other processes.

2. *Generation of foreign substances is not allowed*
   Foreign substances generated due to improper operation and maintenance of equipment, as listed here, must be reduced to Zero by innovative countermeasures.
   - Foreign substances generated from equipment itself, such as leaked lubricants, hydraulic oils, cutting fluids, solvent mists, cooling water, and abraded metal powder from sliding parts
   - Workpieces dropped from equipment
   - Fluid leaked from pipes
3. *Generated by factors other than equipment*
   Thorough countermeasures against the following foreign substances are taken in Step 6 when operators' activities are expanded throughout the entire process. It might, however, be desirable to take this type of countermeasure earlier if possible.
   - Dirt, thread, and lint from personnel, damp rags, and work gloves
   - Foreign substances adhering to purchased parts, wrapping paper, and pallet wear
   - Dust, dirt, and insects entering from outside the building

6.4.3 Countermeasures to difficult cleaning areas

Countermeasures against difficult work areas, such as routine cleaning, lubricating, and inspecting tasks, are taken in Step 2, Step 3, and Step 4 respectively. During Step 1, many difficult cleaning areas are detected and recorded in the difficult work area list, as noted here:

- Although equipment requiring periodical cleaning is installed, neither platforms nor handrails are provided.
- Superfluous safety cover is installed. Excessive cleaning time is spent simply opening and closing the safety cover.
- Equipment is designed without careful consideration of cleaning, lubrication, and inspection. Operators neither enter equipment nor view important areas.

- Pipes, hoses, cables, and wires are congested on the floor or around equipment.

In response to these difficult cleaning areas, the following restorative or remedial actions might be necessary:

- Move equipment to a safer place, or provide platforms or handrails.
- Modify the existing safety cover to be able to open and close it quickly.
- Provide an inspection hole, or make essential component parts of equipment removable.
- Install cable trays or pipe supports to tidy up.

When the above actions against difficult cleaning areas are taken, it is essential that the operators themselves make as many of the improvements as possible to the problems which they have detected in their equipment. If they are not able to accomplish it all by themselves due to the involvement of electrical or welding work, then necessary work orders must be sent to the maintenance department. Having operators look for problems in their environment and take appropriate remedial actions with the support of maintenance personnel are effective ways to develop and maintain optimal operating conditions.

6.4.4 Preexamination and postevaluation of countermeasures

Restorative or remedial actions to existing equipment are almost limitless if people approach them without the appropriate criteria. Presumably cost-effective plans sometimes are submitted without careful consideration of technical and monetary matters, and despite the fact that few benefits are gained by reducing cleaning time. To prevent such a situation, keypoints of the problem, locations of difficult areas, targets, details, costs, and benefits all must be examined in advance.

Furthermore, managers must instruct operators to take remedial actions which provide concrete and visible benefits. If, on the other hand, operators are permitted to be satisfied with superficial results stemming from ambiguous strategies, they will never gain the technical skills they need to solve problems in a significant and consistent way.

All frontline personnel must develop the habit of conducting a clear preexamination and postevaluation of any plan in order to always pay attention to cost. To do this, managers might need to establish clear criteria to evaluate the difficulties in cleaning tasks according to working conditions on the shopfloor. In keeping with the given criteria, operators decide what to improve, how to do it, and how much time it will take.

Unfortunately, in the production as well as in the plant engineering and maintenance departments, some managers and engineers lack an appreciation of costs. Beneficial results, therefore, must be provided in terms of safety, quality, breakdowns, minor stoppages, setup and adjustment, and maintainability of equipment.

In support of activities within a developing autonomous maintenance program, managers must always be certain that operators have a proper comprehension of improvement from early on. When this is the case, operators' ideas sometimes result in unexpected and remarkable effects that could never have been achieved by the engineers' traditional approach.

6.4.5 Where-where analysis and why-why analysis

It used to be almost impossible for conventional problem-solving techniques to eliminate chronic losses that resulted from multiple or complex causes. To surmount such difficulties, TPM utilizes techniques such as PM analysis (techniques to analyze the relationship between phenomena and physical causes) and Failure Mode and Effects Analysis (FMEA). These methods make remarkable contributions in solving problems which occur seldom, but for which the discovery of adequate answers in terms of causes is extremely difficult and, therefore, chronically neglected.

Nevertheless, it is absolutely out of the question to expect frontline production and maintenance personnel to conduct PM analysis immediately because even experienced engineers need a certain period of time and practice to be able to deal with it. Moreover, PM analysis and FMEA do not lend themselves to the analyses of continuously occurring, never-ending phenomena, such as the generation of contaminants. The application of PM analysis and FMEA to these phenomena requires too much time and labor.

Operators, therefore, must use simpler and easier problem-solving techniques, such as where-where analysis and why-why analysis, as summarized in Fig. 6.1. In learning how to observe phenomena, followed by planning and taking remedial actions to solve problems, operators first deal with a familiar problem like cleaning. When the source of contamination is successfully eliminated, they feel pleasure and satisfaction with their achievement. The two aims of Step 2 from both the equipment and human perspectives are attained simultaneously.

Where-where analysis and why-why analysis are conducted in sequence, as described here:

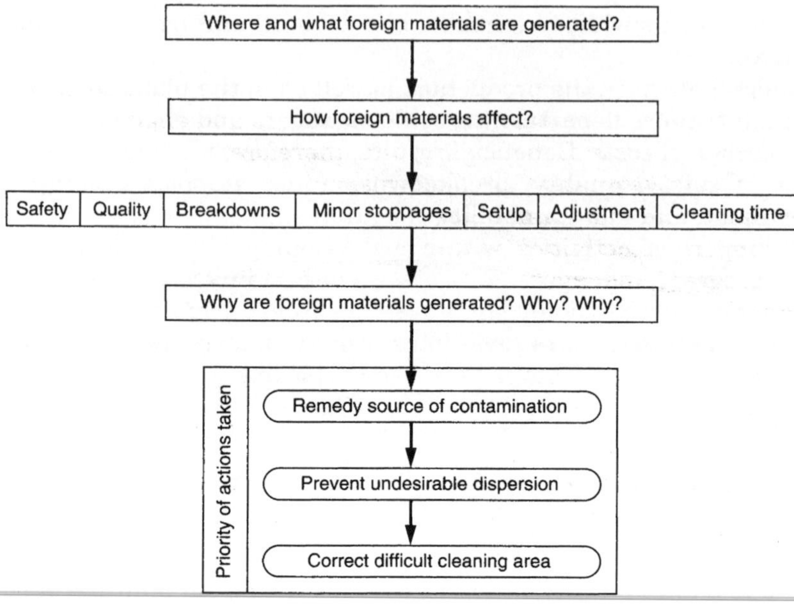

Figure 6.1 A why-why analysis.

1. Each PM group selects one suitable problem occurring in given equipment.
2. It sets goals for improvement and a completion date. To prevent the plan from collapsing, the PM group may set rigorous goals, such as Zero Contamination, which is not ordinarily thought of as a realistic goal.
3. The PM group prepares action plans and schedules by allocating necessary tasks to each operator.
4. No operator should anticipate a solution in advance that is founded on fixed ideas and a traditional way of thinking. After observing relevant phenomena in detail, acquired data must be analyzed based on the repetition of trial and error.
5. Even though one solution may be suggested, no easy countermeasure in response to the problem must be taken to the exclusion of other solutions which may be forthcoming. The aim here is not only effective results, but also education for enabling operators to eventually resolve problems by themselves.
6. If managers find, in the autonomous maintenance audit, that their PM group set overly demanding goals, completion of the goals is allowed by postponing them until the next step. Operators' activity must be evaluated in terms of how much they have learned, rather than by what they have accomplished.

6.4.6 Use corrugated paper and transparent plastic board to prevent dispersion

Measures to prevent the dispersion of foreign substances are developed by PM groups on a trial-and-error basis. At that time, it might be helpful to remove an existing cover if it is not dangerous to do so. Instead of it, operators fabricate a temporary cover with corrugated paper, transparent plastic board, or any other suitable material.

Even these materials are strong enough to stand up to a short period of usage. Operators can easily try various shapes of covers and methods of installation. In addition, no particular skill like welding is necessary, as in the case of a metal cover, and the transparent board makes the outflow of foreign substances visible. After arriving at a satisfactory solution, the operators then request the maintenance department to produce a modified permanent cover made of the same, or even better, material than the original.

6.4.7 Keep records of equipment modifications

Any modifications of equipment, even minor ones, must be recorded as written documents. Handwritten memoranda, or even simple sketches which operators produce after minimal training, are acceptable. Subsequently, engineers, who are selected from the maintenance or plant engineering departments, must check the modified area on-site based on the operators' information and then revise the master drawings and related documents. These same engineers may be able to deal with MP information simultaneously.

It is the drawings that function as the foundation of all technical information. In many companies, such important information differs from actual equipment conditions. Sometimes, it is unclear who modified the equipment, when it was modified, and for what purposes. Under such circumstances, satisfactory facility management is never realized. For example, a piece of equipment was unexpectedly repaired in a factory by a subcontractor using old, unrevised drawings in response to a condition that existed before subsequent improvements were made. In this situation, even the in-house maintenance department would likely make the same mistake.

6.4.8 Review unresolved issues

PM groups might have certain unresolved issues during stepwise activities stemming from various understandable reasons. For example, operators are challenging Zero Contamination, but the ordered parts are not available yet due to a vendor's delivery schedule, or equipment cannot be shut down because of the production plan, and so on.

At the end of Step 2, operators must assess these residual issues and their progress, and revise the schedule to resolve them in Step 3. In the autonomous maintenance audit, managers must carefully check these issues and give operators suitable advice.

6.5 Case Study

In Step 2, operators take various restorative and remedial actions to sources of contamination and to difficult cleaning areas. Several examples are examined in this section. The remarkably effective results described are never attained by means of the traditional approach to improvement (or "Kaizen"). Because operators make great efforts in Step 1, and understand what and where the sources of contamination are, they definitely now want to avoid additional contamination of equipment, and the resulting tough, routine cleaning.

Managers first help operators understand the reason why improvement is needed by pinpointing targets in the context of specific topics, such as sources of contamination and difficult cleaning areas, and then by teaching them how to achieve improvement. This type of TPM approach to improvement is the key to realizing successful results, as described next.

6.5.1 Preventing chips from scattering: the "coverless cover"

In a machinery manufacturing plant, when operators cleaned equipment thoroughly during Step 1, they found that some control panels that were supposed to be sealed up were full of metal chips. In the light of this incredible fact, they decided, as a general rule, not to cover a source of contamination when it is a case of preventive measures against scattering chips. The following example is a typical application of this rule.

In a machining process of a cam shaft, chips scattering in and around machine tools badly affected operating conditions. To begin with, operators spread a white cloth on the floor to carefully observe the motion and distribution of chips, as illustrated in Fig. 6.2.

According to this observation, it was obvious that 11 out of 155 chips scattered during the machining of a cam shaft. Those 11 chips fell to the floor, 8 chips within a radius of 1 meter from the cutting tool, and the other 3 chips within several meters. The former 8 chips are explained by a calculation of forces associated with the rotating speed of the workpiece. The latter 3 chips, however, were influenced by some additional force (see observation (1) in Fig. 6.2).

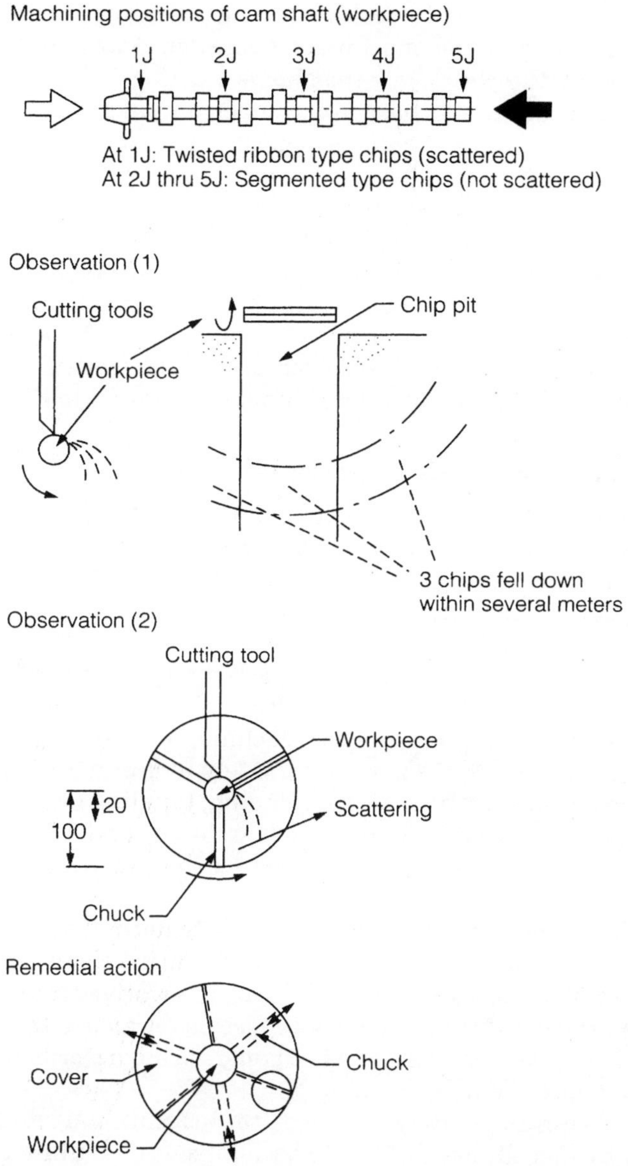

Figure 6.2 The "coverless cover" in a machining process of cam shafts.

To discover any clues to resolve this matter, operators watched a slow-motion video of the machining operation and found that chips were flipped by the chuck jaw that holds the workpiece. By a calculation made by an engineer based on the peripheral speed of the chuck, the distribution of these 3 chips is logically explained (see observation (2) in Fig. 6.2).

Based on the above observation, operators created a "coverless cover" which was conceived from an idea taken from the iris diaphragm of a camera. When a workpiece is set onto machine tools, the chuck is covered automatically by thin steel plates (see the remedial action in Fig. 6.2) which appears as if no cover is installed there. As a result, some 15,000 scattering chips daily are completely controlled and collected into a chip pit. The time spent for cleaning is dropped to Zero.

6.5.2 Countermeasure to drilling chips

As illustrated in Fig. 6.3, workpieces move from the No. 1 to the No. 5 drill to be machined by an exclusive multispindle drilling machine enclosed by a large safety door. This door was installed at the side of a walkway for safety purposes, but it obstructed the operators' easy access to an essential area.

This machine, however, generated a large quantity of 300- to 350-mm long, wiry chips. Because these chips, along with the cutting fluid, were dispersed in and around the machine, operators needed to stop the machine and clean it for approximately 5 minutes 13 times a day. In addition to these problems with cleaning, tangling long chips caused frequent drill breakages, limit-switch failures, and even damage to the sliding surface of the machine bed as the precision drilling tools reciprocated.

Such conditions once were considered quite normal and no one questioned the time and effort involved in the frequent cleaning. Operators decided, in this instance, to take an innovative countermeasure founded on the breakthrough approach, so as to remove the safety door and eventually be able to satisfy the general rule described in the previous case study.

In the beginning, they applied the traditional and ordinary approach by installing minimal covers in front of drilling tools. Damage to the sliding surface was prevented, but the drilling tool ports of newly installed covers were abraded instantly by the long wiry chips. Operators then executed several other ideas on a trial and error basis to shorten the long chips without lowering the machine speed. Finally, they installed, on trial, an unused cutting tool at the opening of the cover, and thereby succeeded in shortening chips to 60 to 70 mm (see remedial action (1) in Fig. 6.3). Thereafter, the wiry chips

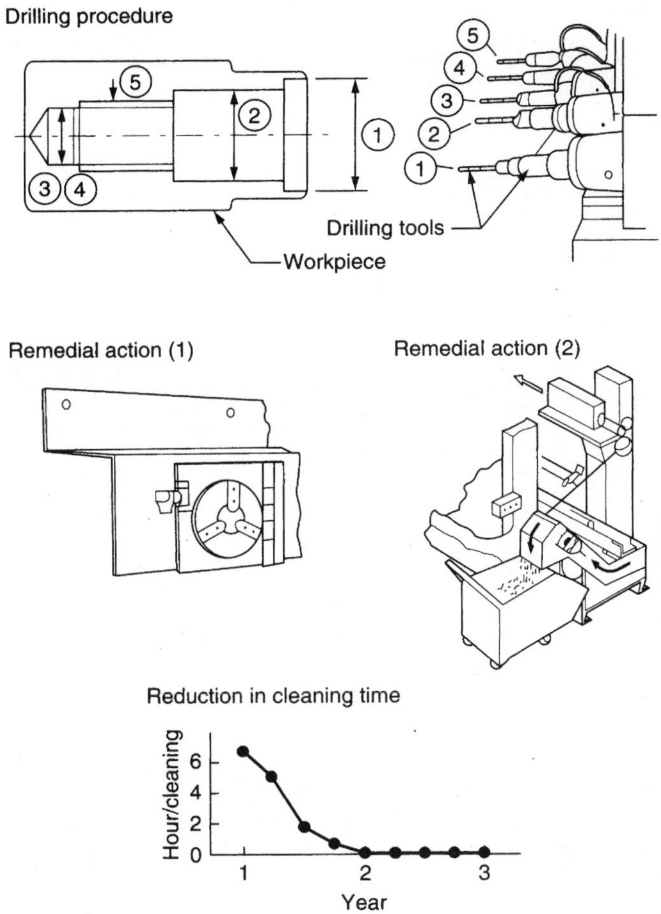

Figure 6.3 Remedial actions taken against long, wiry chips.

dropped into a chip pan as expected. The safety door, as a result, was removed.

Nonetheless, operating interruptions and machine downtime for cleaning continued to occur because the quantity of chips itself was never reduced. Operators, eventually, modified the chip pan so as to wash chips to the outside of the machine by using circulating cutting fluid. Moreover, they invented a chip conveyor without a motor, but driven, instead, by the drilling machine itself (see remedial action (2) in Fig. 6.3).

By means of the above countermeasures, the chips fell without operators' assistance into a chip carrier that needed to be emptied only once a day. Moreover, operators now cleaned their machine in 10 minutes a

From where (source)	What contaminant	What kind of effects	Why (causes)
Bearings assembled in a jig in a vacuum evaporation system	Abraded metal powder	[Quality] Powder adhered to evaporation material results in undesirable changes in quality of thin film and ΔVFB. [Equipment] In case of deviation from specified value in ΔVFB, equipment operations are stopped.	Why is powder generated? ↓ Bearings made of metal are abraded due to friction. ↓ Why are bearings abraded? ↓ General-use mass-produced bearings are utilized without applying lubricant because of extreme vacuum pressure. ↓ Why are vacuum-use bearings not utilized? ↓ Those bearings which satisfy requirements in terms of dimension along with operating conditions under high temperature and vacuum pressure were not available in the market when process was engineered. ↓ Are vacuum-use bearings not available now in the market?

Figure 6.4 The why-why analysis applied to a thin film deposition system.

Corrective actions				
Sources of contamination	Dispersion prevention	Periodic inspection	Equipment	Feedback
(1) Examine availability of high temperature and vacuum-use bearings. (2) Install wider gap type bearings.	(1) Examine availability of sealed type bearings.	(1) Continuously follow the same inspection procedures whenever equipment is cleaned.	—	Issue MP information sheets.

As results of above assessment:

(1) Replace, on trial basis, existing bearings with vacuum-use bearings (new standard model) assembled upstream to evaporation source which highly affects ΔVFB.

(2) For those locations except above, replace existing bearings with wider gap type, and evaluate availability and lifetime.

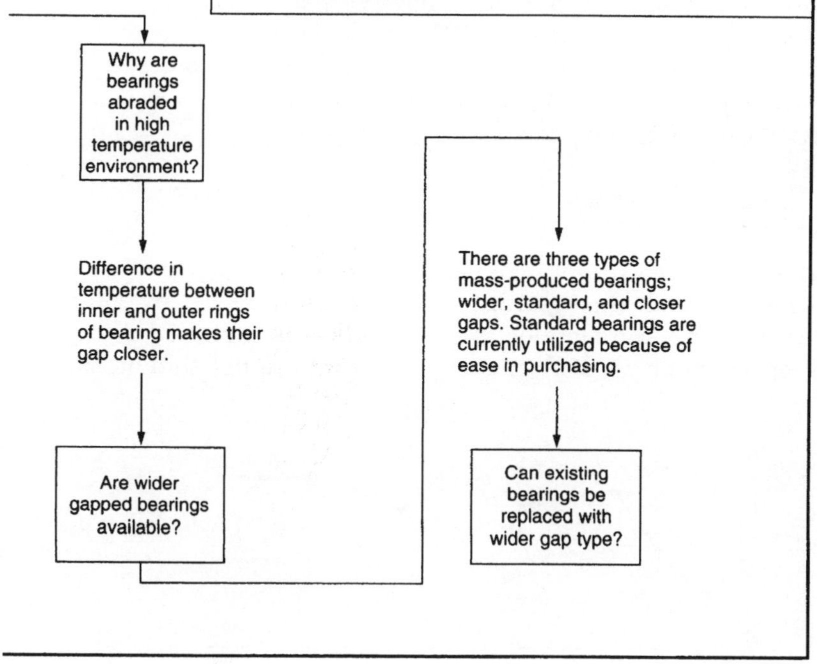

Figure 6.4 *(Continued)*

day in order to maintain the cleanliness attained in Step 1, whereas it previously required some 6 hours. Machine downtime for removal of chips went from 65 minutes to Zero.

6.5.3 Eliminating foreign particles in semiconductor manufacturing vacuum processes

In the semiconductor manufacturing process, cleaning is a very important role in the quality and yield of products. In the vacuum process, such as an ion implanter and sputtering system, setup work is nothing but cleaning. When the consequence of these preparations does not satisfy specified inspection criteria, operators must open the equipment to clean up inside again. Because of these repetitive cleanups and subsequent air evacuations, large time losses occur in producing the required vacuum pressure.

Figure 6.4 is an example of why-why analysis applied to troubles caused by abraded metal powder leaking from a bearing installed in a vacuum process. In such a process, lubrication is taboo. Accordingly, various problems arise. As a result, however, of taking remedial actions involving why-why analysis, the time needed to clean the equipment is substantially reduced, and cleanliness becomes completely satisfactory. As of today, operators are maintaining these conditions, as shown in Fig. 6.5.

6.5.4 Measures for containing emery powder

Window glass installed in automobiles must have a logogram which indicates that it meets the relevant safety regulations set by countries of destination. Automobiles in which glass is installed either without a logogram or with an unclear logogram are not sold on the market.

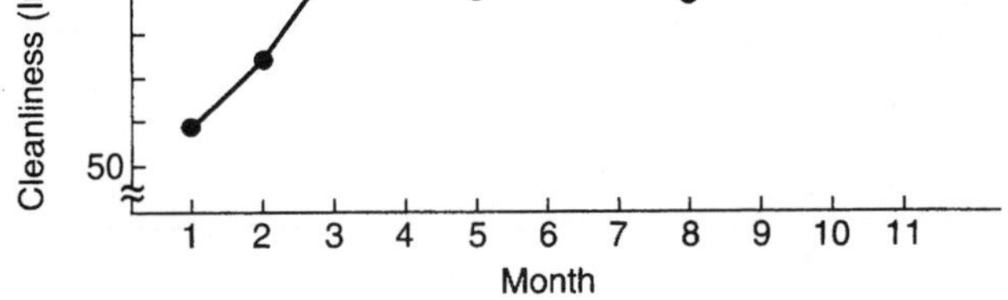

Figure 6.5 Cleanliness achieved in vacuum evaporation systems.

In Japan, this logogram is etched by a blast of emery powder onto a stencil applied to the glass. The powder that blows out from the blaster and stencil used to have a variety of harmful effects on quality and equipment. A PM group, therefore, initiated a series of remedies after deciding on the following targets, schedule, and each operator's role.

- What: Powder leakage
- Goal: Reduce to Zero
- How long: Five months from May to September

Operators developed remedial actions on a trial and error basis. Their progress is outlined in this section.

1. *Identify the status quo by where-where analysis.*
   Until then, a long time was needed to clean the blaster and the surrounding floor free from the powder leaked at the gap between the glass and the sealing ring of the marker and at the flanged connections of the blaster. As a result of the former leakage, air cylinders and bearings in pillow blocks were damaged. By the latter leakage, not only chains and sprockets were damaged, but also glass was scratched by powder adhering to transfer belts.

   Operators decided to perform a where-where analysis first. They started a survey to identify from where, how, and to what extent the powder was leaking. To these ends, the blaster and marker were divided into upper, middle, and lower sections with corrugated paper, as illustrated in Fig. 6.6. From this survey, data on leaking powder was obtained, as shown in Fig. 6.7 (see before improvement).

2. *Analyze causes by why-why analysis.*
   Operators subsequently conducted a why-why analysis based on data obtained in the previous survey. Their assumptions are summarized in Fig. 6.8.

3. *Observe phenomena carefully.*
   They carefully observed the working mechanisms of the machine in order to confirm their inferences obtained by why-why analysis, as illustrated in Fig. 6.9. In discussing how to take remedial actions, they sometimes examined operating conditions of the machine, surveying it with their hands. In this way, another point of information was discovered; the bellows built into the expansion joint of the stencil holder was breathing and resulted in powder leakage.

4. *Hold study meetings.*

   One-hour study meetings were held four times and were participated in by all members. They called in their manager, and the maintenance and vendor engineers as advisors, to understand the basic structure and function of the machine involved. After having responded to and solved some fifteen questions, they were able to plan and implement remedial actions with full confidence of success.

5. *Execute countermeasures.*

   Cause A

   - Phenomena: Powder adhering to sliding parts caused irregular motion of air cylinders installed to reciprocate upward.
   - Remedial Action 1: Air cylinders were inverted to reciprocate downward so as to be free from powder adherence.
   - Remedial Action 2: Stroke of air cylinder was changed from 15–16 mm to 5–6 mm to minimize powder adherence.

   Cause B

   - Phenomena: Powder leaked from a gap between the glass and the existing, excessively narrow sealing ring.
   - Remedial action 1: An additional "O" ring was applied to the existing metal ring, but powder leakage persisted.
   - Remedial action 2: A rubber sheet gasket was applied. Powder leakage was reduced to Zero, but the installation and removal of the stencil was not easy.
   - Remedial action 3: The sealing ring unit was modified to be removable. Quick stencil changes were achieved.

   Cause C

   - Phenomena: Excessively high blasting pressure caused powder leakage.
   - Remedial action: Making a comparison with the given logogram samples, trial blasts were carried out under different conditions of blasting pressure and time. A specified quality was obtained by using a lower pressure than the original.

   Cause D

   - Phenomena: Flanged connections tightened with nuts and bolts without a gasket resulted in powder leakage.

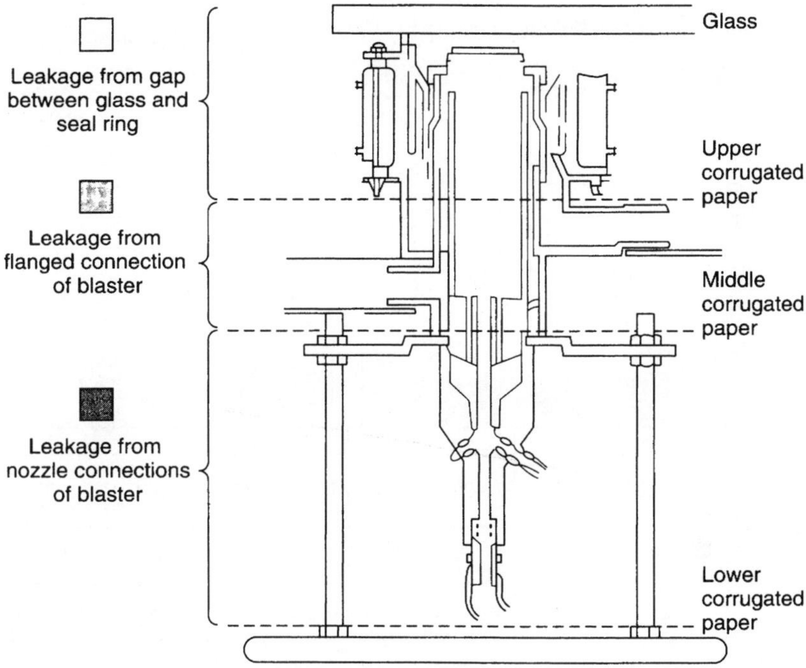

Figure 6.6 Observing the leakage of emery powder.

- Remedial action 1: Nuts and bolts were removed and a gasket was installed, but powder leakage was not reduced.
- Remedial action 2: A gasket was bored with bolt holes and the connections were tightened with nuts and bolts. Powder leakage was reduced, but some still remained.
- Remedial action 3: Another "O" ring was added and leakage disappeared.

Cause E

- Phenomena: Due to the poor capacity of air exhaust, bellows in the expansion joint was breathing.
- Remedial action: Airflow was checked by feeling with hands. Air was exhausted insufficiently from a cyclone built into the powder collector. An existing metal cover installed on the top of the collector was replaced with a sponge filter in order to evacuate the air sufficiently.

Cause F

- Phenomena: Airflow was checked by feeling with hands. Air was drawn in from an exhaust-hose side during blasting.

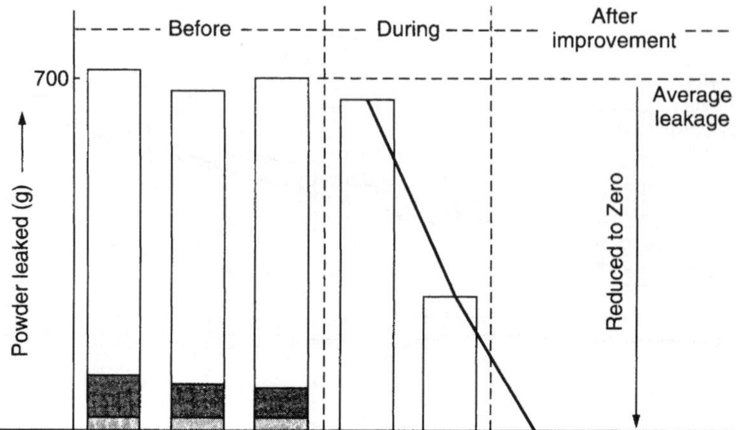

Observation	1	2	3	1	2	1	2	3
Number of glasses blasted	13,523	12,911	13,002	13,125	13,241	13,540	14,012	13,871
Powder spent (kg)	94.66	90.37	91.55	91.87	92.68	92.35	95.97	95.00
Date observed	5/7~9	5/16~18	5/23~25	8/10~12	8/22~24	9/21~23	9/24~26	9/27~29

Figure 6.7 A reduction in the leakage of emery powder.

- Remedial action: A check valve was installed at the outlet of the marker. Breathing of bellows in expansion joint was prevented by eliminating backward airflow.

6. *Execute additional countermeasures.*
   By the full implementation of countermeasures with where-where and why-why analyses, powder leakage from equipment was completely eliminated. Operators, however, continued to find a slight amount of powder remaining on the stencil. They, therefore, decided to change the timing for the start of the descent of the marker. As the result of repetitive experiments whereby the machine setting for the start of descent was changed from 1.0 to 1.5 seconds after blasting, all residual powder finally disappeared.
7. *Confirm effective results.*
   Operators conducted the same survey again to confirm the elimination of powder leakage. The survey indicated that leakage had been reduced to Zero, as illustrated in Fig. 6.7 (see after improvement).
8. *Prevent recurrence of problems.*
   To maintain revised operating conditions, routine control items and inspection methods are compiled onto check sheets.

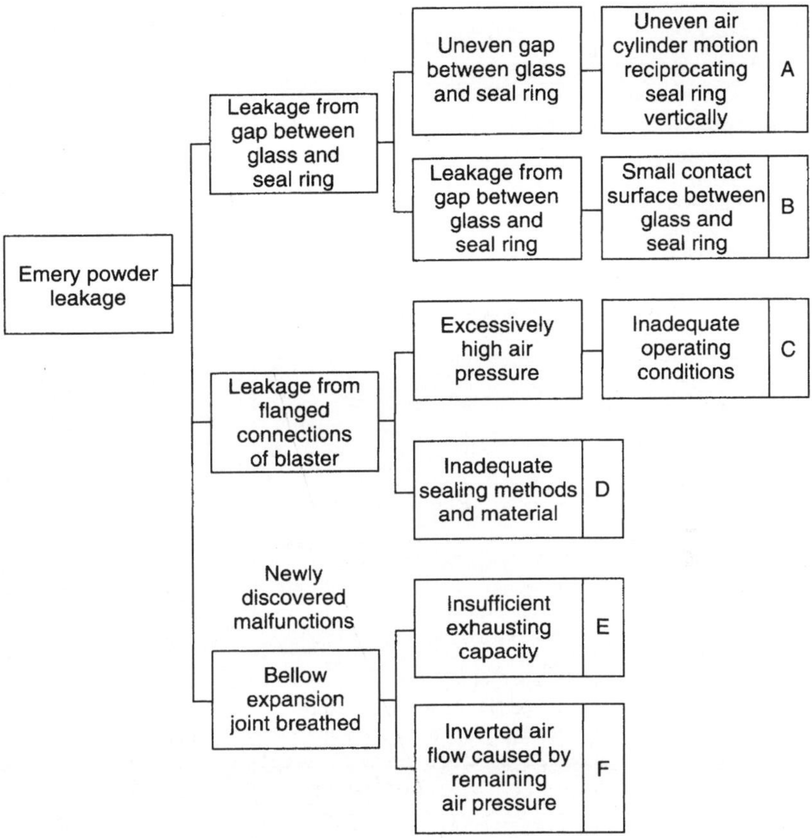

Figure 6.8 The why-why analysis applied to the leakage of emery powder.

6.5.5 Cleaning difficulties at a dipping basin

For the purpose of preventing adherence, rubber sheets were dipped into a corrosive chemical liquid transferred by a net conveyor, as illustrated in Fig. 6.10 (see before improvement). Operators needed to spend a long time cleaning this equipment. In addition, lubrication and inspection were impossible because the conveyor and shafts were installed in the dipping basin. As might be expected, the corrosive liquid damaged the shafts and the agitating air pipes. Air pipe clogging also caused frequent, minor stoppages of the entire process.

In response to this situation, the PM group in charge decided to address this problem as a subject of Step 2. Operators went on to improve equipment, as summarized in Fig. 6.10 (see after improvement).

	Causes	On-site observations	Remedial actions planned
A	Uneven air cylinder motion reciprocating seal ring vertically	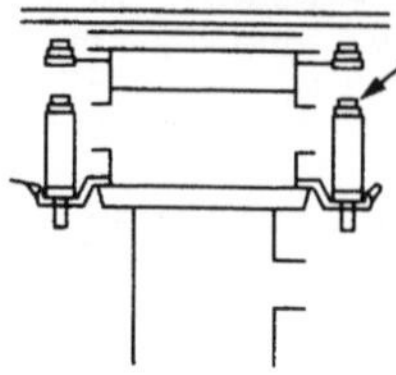 Leaked powder inter-mingling in air cylinders	Invert air cylinder installation downward in order to minimize adherence of powder onto cylinders rod.
B	Small contact surface between glass and seal ring	Excessively narrow seal ring; Stencil 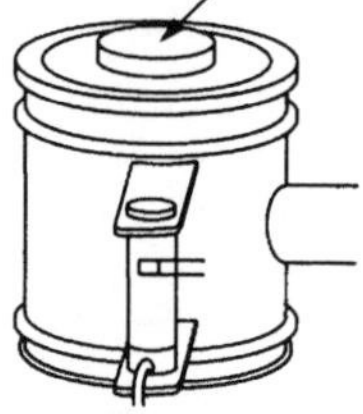	Widen seal ring and devise gasket applied on cylindrical metal cover.
C	Inadequate blasting air pressure	Air pressure and legibility of logogram • Compressed air Supply side: 4.0 kg/cm^2 Blasting side: 3.5 kg/cm^2 • Blasting time: 0.8 sec • Legibility of logogram: More legible than given sample	To etch the most legible logogram with minimum dispersion of leaked powder, optimum blasting pressure must be determined by changing operating conditions.
D	Inadequate sealing methods and material	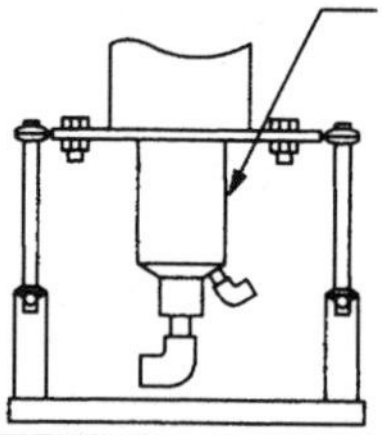 Flanged connection of blaster is tightened with nuts and bolts, without gaskets	Insert gaskets.

Figure 6.9 Considering remedial actions by way of careful observations.

Table 6.2 lists the remarkable results of the actions taken. The occurrence of minor stoppages was reduced from 20 to Zero times per month.

6.6 The Keypoints of an Autonomous Maintenance Audit

First of all, auditors must carefully check that all operators are familiar with group activities, and that autonomous maintenance activities are well implemented by each PM group. Step 1 may be developed involving only physical cleaning work. Activities from now on, how-

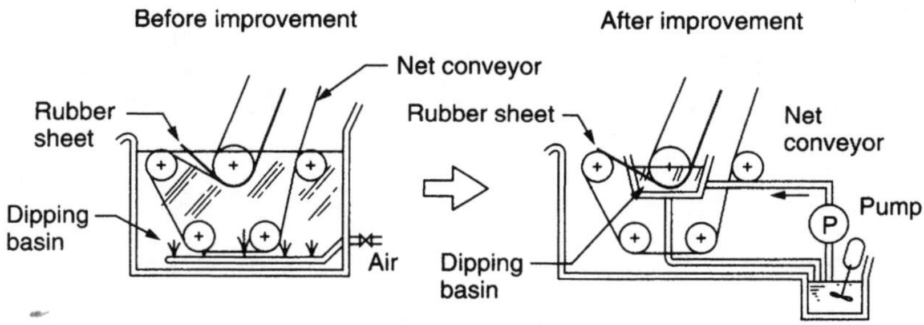

Figure 6.10 Improving a difficult cleaning area.

ever, are more an exercise and practice of brain power than muscle power. If operators do not adequately recognize the deterioration and defects of equipment, they neither understand "cleaning is inspection" nor aggressively strive for improvement.

In the audit, it is more important to confirm the following points than whether or not cleaning time targets were met.

- What did operators learn about matters such as structure and function of equipment, influences of contaminants, and so on?
- How thoroughly did operators survey for harmful effects and related causes?
- How substantial were the operators' efforts by way of trial and error?
- How effective were the operators' countermeasures in terms of safety, quality, breakdowns, minor stoppages, setup, and adjustment?

If the above checkpoints are not satisfied, the PM group involved is judged as having failed in spite of its having completed cleaning within the given time target. The true aim of the activities of Step 2 exists in developing knowledgeable operators who resolve problems by themselves.

When a PM group insists on targeting lofty goals, but has not attained them yet, the group must clearly describe to the auditors a

TABLE 6.2 Successful Results of Remedial Actions Taken

Criteria	Before	After
Spent time (minutes/cleaning)	180	15
Net conveyor life (month)	4	36
Minor stoppages/month	20	0

TABLE 6.3

Step 2	Autonomous Maintenance Audit Sheets	Sheet 1 of 3			
No.	Audit points	Results			
1.	**Step 1 conditions**				
1.1	Is cleanliness attained in Step 1 well maintained?				
1.2	Are residual issues left over from Step 1 resolved?				
2.	**Group activity** (General)				
2.1	Are aims of Step 2 understood adequately?				
2.2	Is activity plan made in advance? Well executed?				
2.3	Are managers' models well understood?				
2.4	Is activity board adequately utilized?				
2.5	Are defective areas accurately located? Confirmed by all members?				
2.6	No overlooked defective areas detectable with current knowledge?				
2.7	Are identification tags collected and kept after remedies?				
2.8	Are questions sufficiently posed?				
2.9	Are questions definitely resolved? Understood by all members?				
2.10	Is activity developed in following priority?				
(1)	Remove generation of contaminants at source.				
(2)	Prevent contaminants from dispersing.				
(3)	Improve cleaning methods and tools.				
(4)	Modify equipment to make cleaning tasks easier.				
2.11	Are safety matters carefully respected?				
2.12	Are TPM activity hours and frequency adequate?				
2.13	Is more efficient way of TPM activity pursued?				
2.14	Are used spare parts and consumables recorded?				
2.15	Is meeting after on-site activity definitely held? Reports submitted?				
2.16	Is activity participated in by all members?				
2.17	Are all members cooperating equally? Not led by particular member?				
2.18	Are noteworthy ideas introduced actively to other PM groups?				

TABLE 6.3 *(Continued)*

Step 2	Autonomous Maintenance Audit Sheets	Sheet 2 of 3			
2.19	Is cooperation with full-time maintenance satisfactory?				
3.	**Tentative cleaning standards**				
3.1	Are cleaning time targets, work allocation and areas clearly specified by manager? Are these prescriptions well understood?				
3.2	Are tentative cleaning standards set at the beginning of Step 2?				
3.3	Are cleaning intervals identified to maintain cleanliness achieved in Step 1?				
3.4	Are tentative cleaning standards revised as every remedial action is taken?				
3.5	Is cleaning work to be done during up/downtime of equipment clearly distinguished? Well understood?				
3.6	Can everyone clean equipment in accordance with cleaning standards? Time targets achieved? If not, are adequate plans and schedules prepared?				
4.	**Sources of contamination**				
4.1	Are remedial actions taken deliberately, based on detailed assessment of actual generating conditions of contaminants?				
4.2	Are locations and manners of contaminants clearly recognized by detailed observation? Where-where analysis conducted, if necessary?				
4.3	Are structure and function of equipment sufficiently learned? Why-why analysis conducted, if necessary?				
4.4	Are all simple sources resulting from deteriorated parts such as water, oil and steam leakage remedied?				
4.5	Are cost and effects of remedies reviewed? Actual figures recorded?				
5.	**Difficult cleaning area**				
5.1	Are remedial actions taken deliberately, based on detailed assessment of actual conditions of cleaning tasks in awkward work areas?				
5.2	Are sources thoroughly remedied to make actions against difficult cleaning areas unnecessary?				
5.3	Are costs and effects of remedies reviewed? Actual figures recorded?				
6.	**Equipment** (Main body and surroundings)				

TABLE 6.3 *(Continued)*

Step 2	Autonomous Maintenance Audit Sheets	Sheet	3 of	3	
6.1	Are tools and jigs stored at designated locations? No shortage or damage?				
6.2	No unnecessary pieces of equipment?				
6.3	No unnecessary materials in process? No dropped parts on floor?				
6.4	Can quality/defective products or scraps be clearly distinguished?				
6.5	Do status display and warning lamps function properly?				
6.6	Do safety devices function properly?				
7.	**Lubrication**				
7.1	No dirt, dust or leakage in and around air lubricators, oil reservoirs or centralized systems?				
7.2	Is oil level proper? Adequate lubricants at lubricating surfaces?				
7.3	Are lubricants not contaminated? Not deteriorated?				
8.	**Visual control**				
8.1	Are visual controls devised to facilitate cleaning tasks?				
8.2	Are available visual controls installed using current technical knowledge and skill?				
9.	**Short remedial program**				
9.1	Is subject selected from the six big losses?				
9.2	Are problems clearly identified? Targets pinpointed?				
9.3	No easy countermeasures taken?				
9.4	Are cost and effects of program reviewed? Actual figures recorded?				
9.5	Are preventive measures against recurrence of problems provided?				
10.	**Residual issues**				
10.1	No residual questions, equipment defects? Are secure plans and schedules for corrective actions prepared?				
10.2	No residual remedies against sources of contamination and difficult cleaning areas? If so, do reasonable explanations exist? Are plans and schedules to solve these issues clear?				

summary of unresolved issues, an action plan, and a schedule to undertake additional efforts in the next step. Managers must conduct question-and-answer sessions to help operators see the future direction of autonomous maintenance activities rather than focus on excessive details of techniques for improvement.

Not to be overlooked is the managers' need to pay careful attention to preventing the dropout of PM groups from the TPM program, which has the highest probability of occurrence in this step. From this point of view, the actual auditees who are evaluated in this audit are the frontline managers and supervisors, who must manage and train operators through routine work. An example of audit sheets is shown in Table 6.3.

Chapter

7

Step 3: Cleaning and Lubricating Standards

7.1 Aims from the Equipment Perspective

In Step 1, operators thoroughly remove contamination neglected and accumulated over a long time. The sources of contamination and difficult cleaning areas which impede the upkeep of newly achieved cleanliness are remedied in Step 2. Equipment thereby becomes so clean as to appear almost new, and this condition is maintained easily by following the cleaning standards set by operators themselves.

In Step 3, an overall inspection of lubricating points and surfaces takes place in order to identify and remedy the defective areas stemming from lack of lubrication, especially in difficult lubricating areas. By way of the restorative and remedial actions resulting from this inspection, proper and reliable lubrication is maintained. Operators thereby finally achieve higher reliability and maintainability of equipment by faithful cleaning and lubricating. Step 3 has the most important role in completing the establishment of the basic equipment conditions.

7.2 Aims from the Human Perspective

During Step 3, operators set cleaning and lubrication standards in order to maintain achieved equipment conditions. Prior to the initiation of TPM, these rules were prepared by staff personnel from the plant engineering, maintenance, or production departments. Then, managers would assign these rules to operators and force their implementation. This has been a common reality for factory management in many companies.

In TPM, however, which aims at autonomous supervision, these rules are set, based on actual experience, by the operators who must follow them. Of course, the rules should not be the products of imagination, as is the case when they are set by staff.

Operators, when setting standards on their own, make them both easy to follow and conducive to favorable operating conditions. They recognize the necessity and importance of following the standards. In these circumstances, the operators are certainly inclined to follow their own rules. Furthermore, this approach impresses the operators with the importance of their role in plant operations. This is the first step to autonomous supervision.

7.3 How to Develop Step 3

7.3.1 Maintain the cleanliness achieved in Step 2

Having maintained the cleanliness audited at the end of Step 2, operators must now develop Step 3. Along the way, managers reinforce with the operators the concept that "cleaning is inspection."

There may be some residual issues of remedial actions left over from Step 2 because of overly ambitious goals. Operators should continue to make their best efforts to finish these projects as early as possible. All PM groups are persuaded to finalize all items related to cleaning and lubrication in Step 3.

7.3.2 Cumulative improvements

When operators try to clean and lubricate equipment simultaneously, within a given time target specified by the combined tentative cleaning and lubricating standards, many unforeseen problems appear. These new difficulties stem from the fact that a certain limitation resides in both the distance that operators are able to move and the amount of hand tools and lubricants that they are able to carry about. Some factors that relate to these problems are the operators' work motions and route for lubrication, the manner and location for storing tools and lubricants, and the feasible combination of cleaning and lubricating within designated time targets.

Thorough lubricating sometimes causes dripping or leaking of excess lubricants. As a result, lubricants become a new source of contamination and demand another remedy. Therefore, to finish cleaning and lubricating within given time targets, further patient efforts to improve work methods and equipment must be forthcoming.

7.3.3 How to divide a step into substeps

Operators are taught about lubrication by maintenance engineers at the beginning of Step 3. When they gain basic knowledge and skill, they perform relevant practice in the following sequence:

Check	Identify lubricating points and surfaces. Detect defective areas in equipment related to lubrication.
Act	Remedy defective areas and modify difficult lubricating areas.
Plan	Set cleaning and lubricating standards.
Do	Execute cleaning and lubricating standards.

The above CAPD cycle is repeated until time targets are met. Table 7.1 shows an example of these substeps.

TABLE 7.1 Dividing Step 3 into Substeps

Substep	Major activity
1	Conduct education for lubrication.
2	Identify lubricating points and surfaces.
3	Allocate routine lubricating tasks.
4	Set tentative lubricating standards.
5	Estimate lubricating intervals.
6	Set lubricating time targets.
7	Set improvement targets.
8	Sources of contamination:
8-1	Identify lubrication related sources.
8-2	Plan remedial actions.
8-3	Take remedial actions.
8-4	Evaluate results of actions.
9-1	Identify equipment defects.
9-2	Restore equipment defects.
10	Difficult cleaning areas:
10-1	Conduct a why-why analysis.
10-2	Plan remedial actions.
10-3	Take remedial actions.
10-4	Evaluate results of actions.
11	Review lubricating standards.
12	Compare with lubricating standards set by full-time maintenance.
13	Set cleaning and lubricating standards.
14	Develop a short remedial program.
15	Conduct an autonomous maintenance audit.

7.4 Establish a Lubrication Control System

7.4.1 What is lubrication control?

In spite of hidden, but significant, losses caused by poor lubrication in the past, no one paid adequate attention to lubrication. In some cases, extraordinary lubricating tasks were left to operators and decisions about the types of lubricants always, without question, followed the vendors' instructions. Nor was it unusual that every installation of new machines involved new types and brands of lubricants. Nevertheless, the maintenance department focused on changing and reclaiming contaminated oil at irregular intervals without any systematic policy.

These conditions were representative of the situation that existed in most companies at the time TPM was introduced. No one doubted the existence of such problems as frequent oil leakages, contaminated lubricants, negligence of lubrication, and so on. A factory, for instance, did not realize for a long time that a large quantity of oil leakage was occurring from a cracked reservoir covered with thick dust and greasy sludge.

As defined in Fig. 7.1, lubrication control appears very simple and easy. This might be one of the reasons why lubrication control was not

Lubrication control

1. Supply lubricant which is clean
2. Conforming to operating conditions
3. Through designated lubricating points
4. To designated lubricating surfaces
5. At designated time
6. In proper quantity

Lubricating point and surface

Lubricating point: Point at which to supply lubricant

Lubricating surface: Sliding parts which require lubrication such as reciprocating and rotating parts, and chains

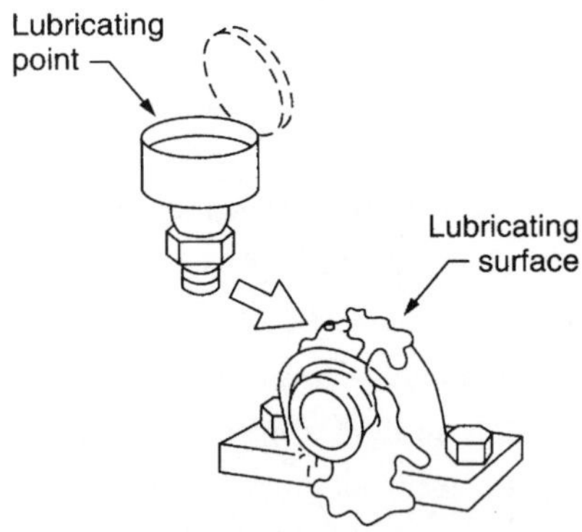

Figure 7.1 Lubrication control.

seriously considered by engineers and relegated to operators and maintenance technicians. In order to establish a firm lubrication control system, the procedure diagrammed in Fig. 7.2 must be conducted and the mutual cooperation of autonomous maintenance and full-time maintenance must be enlisted.

7.4.2 Preparation by the maintenance department

Prior to the commencement of Step 3, the maintenance department must make necessary preparation as follows:

Set lubrication control rules

- Integrate and minimize types and viscosity of lubricants.
- Assign code numbers and identification colors for each type of lubricant, as detailed in Table 7.2. Then, set color control rules.
- Prepare lubrication labels which indicate types and viscosity of lubricants, lubrication intervals, and work allocations. Figure 7.3 shows a typical sample of lubrication labels.
- Specify how to indicate lubricant levels at air lubricators, level gauges installed at the side of reservoirs, and any other types of oilers.
- Specify lubricant control rules, such as the central and local storage of the lubricant, a container, inventory, supply and disposal, and place responsible personnel in charge of each kind of storage.
- Prepare samples and models for demonstration if necessary.

Prepare teaching materials

- Materials to teach the basic theory of lubrication, major rotating and reciprocating parts, plus the structure and function of lubricating tools and apparatus
- Lubrication control manuals
- Lubrication inspection manuals and check sheets
- Audiovisual materials, one-point lessons, and cutaway models

7.4.3 Launch stepwise activities with lubrication education

Lubrication-related education is conducted at the beginning of Step 3. Other detailed education for operators in matters of categorized overall inspection must take place in Step 4. Accordingly, the maintenance department determines the educational policy and its details, and makes necessary preparations in advance, as discussed earlier.

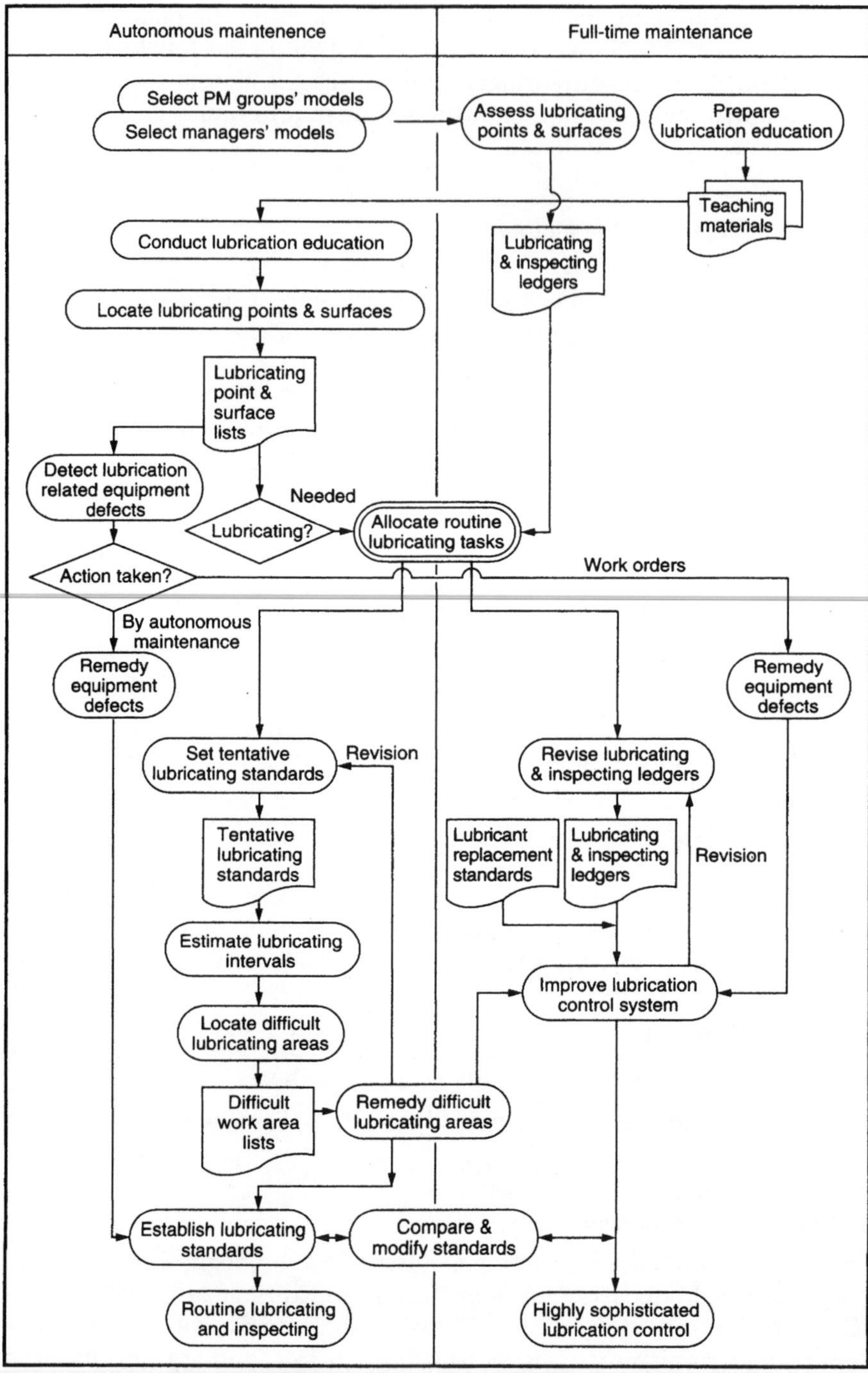

Figure 7.2 Implementing a lubrication control system.

TABLE 7.2 Typical Color Codes for Lubricants

Type	Color code	Classification	Major services
Hydraulic oils	Yellow-32	ISO VG 32	Low-pressure hydraulic system; Air lubricator
	Yellow-56	56	High-pressure hydraulic system
Machine oils	Blue-32	32	Sliding surface for common machines
	Blue-68	68	Sliding surface for particular machines such as numerical control machines and milling machines
Gear oils	Green-6	6	Main shaft bearing for grinding machines
	Green-32	32	Common reduction gears operated higher than 500 RPM
	Green-68	68	Particular gears operated lower than 500 RPM
	Green-150	150	Heavy-duty machines; Worm gear for 15 ton roll forging machines
Compressor oils	Orange-26	26	Screw compressors
	Orange-46	46	Reciprocating compressors
Greases	White-B-1	NLGI No. 1	Sliding parts and bearings for common machines (Centralized greasing systems using long conduits)
	White-B-2	1	Sliding parts and bearings for heavy-duty machines (Centralized greasing systems using long conduits)
	White-C-1	2	Sliding parts and bearings for common machines
	White-C-2	2	Unsealed gears and bearings
	White-C-3	2	Heavy-duty machines 6 ton mill-forging machines

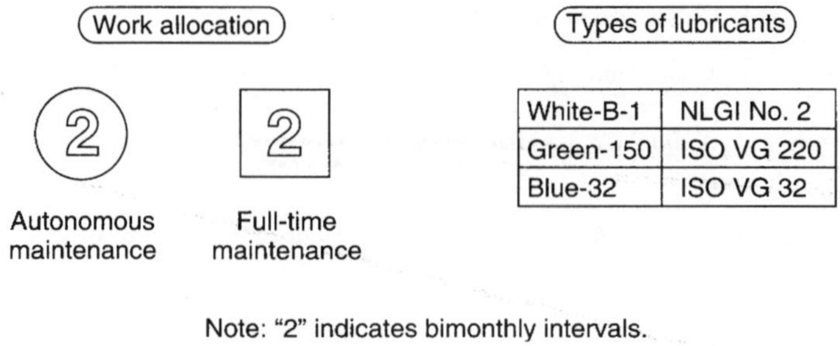

Figure 7.3 Lubrication labels.

The approach to this educational activity for operators with reference to lubrication is the roll-out method. Topics covered are basic knowledge of lubrication, structure of rotating and reciprocating parts, types of lubricants, color lubrication control, lubricating tools, apparatus and methods, inspection items, methods and criteria, lubrication system flowcharts, and case studies of troubles caused by poor lubrication in terms of equipment and quality of products.

7.4.4 Identify lubricating points and surfaces

Operators and maintenance personnel join in pairs and conduct an overall lubrication inspection. All lubricating points and surfaces must be identified by thorough inspection of equipment as well as by a review of related drawings and vendors' manuals. Those points and surfaces that have been unrecognized and neglected, even by maintenance personnel, are often discovered in many pieces of equipment through this procedure. In some cases, lubricating points were painted over and not lubricated for a long time. A detailed survey apparently is required to not overlook such points.

In the course of these inspections, maintenance personnel teach operators how to find deteriorated and defective parts in equipment, and assist with lubricating practice. At the same time, the allocation of routine lubricating tasks is decided upon for each lubricating point and surface. These on-the-job training and practice procedures reinforce operators' skills. In this way, PM groups list the details of lubrication for each piece of equipment, such as lubrication points and surfaces, type of lubricants, intervals, and lubricating and inspection methods.

7.4.5 Allocate routine lubrication tasks

The allocation of routine lubrication tasks is determined by negotiations between the production and maintenance departments. Routine lubrication and inspection, however, are basically performed by operators. The maintenance department, on the other hand, handles those exceptional tasks which require particular skills, out of concern for safety or the necessity to disassemble machinery. Maintenance personnel, moreover, concentrate on the achievement of much higher levels of lubrication control.

The major lubrication and inspection tasks expected to be done by operators are:

- Lubrication by means of hand lubricators and pressure guns, and application with brushes
- Oil level inspection and lubrication at reservoirs and air lubricators
- Temperature and oil level inspection and lubrication at reduction gears, change gears, pumps, compressors, and any other parts as needed
- Inspection of lubricating surfaces (oil quantity, temperature, and clogging) in remote or centralized lubricating and greasing systems

7.4.6 Draw lubrication system flowcharts

Most large machinery has a large number of lubricating points and surfaces and frequently involves a remote or centralized lubrication and greasing system. The operators' load may be reduced if lubricating system flowcharts are included among the vendors' documents. In their absence, it is recommended that operators draw the lubrication system flowcharts by themselves, based on an overall inspection of the system, by checking for lubricating points and surfaces and tracing the connecting lube pipes. Isometric drawings may be easier to understand and are helpful in following routine lubricating and inspecting procedures.

In ordinary processes, the operator's work motions and route for the purpose of lubrication are written onto the plot plan. Some plants call this drawing the "lubricating map." Figure 7.4 shows an example of a lubrication map. If visual control is consistently applied in association with lubrication labels (Fig. 7.3), operators are soon able to lubricate and inspect without these charts. These same operators are not, however, always in charge of the same process. Therefore, these documents should be prepared and made available.

Major questions about inspecting items in remote and centralized lubricating and greasing systems are:

- Is the lubricant supply unit working properly?
- Is the oil level of the reservoir proper? Is the oil pressure adequate?

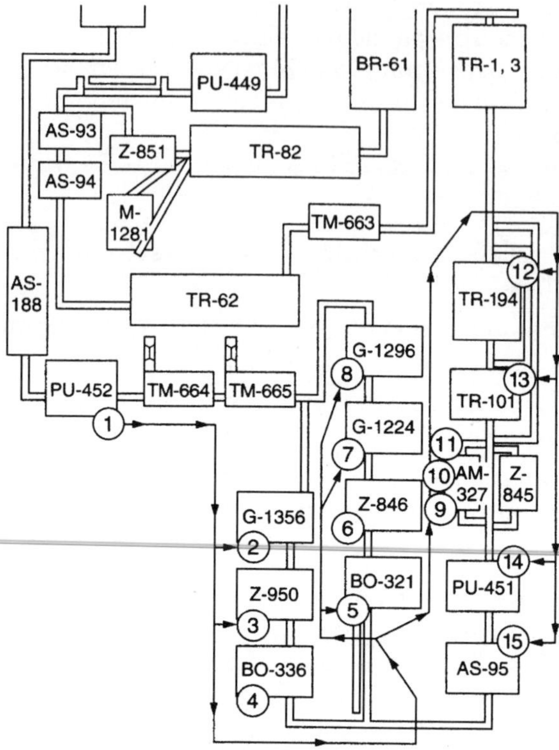

Sequence	No. of lubricating points	Equipment
1	1	PU-452
2	1	G-1356
3	1	Z-950
4	1	BO-336
5	1	BO-321
6	1	Z-846
7	1	G-1224
8	1	G-1296
9	1	AM-327
10	1	ditto
11	1	ditto
12	1	TR-194
13	1	TR-101
14	1	PU-451
15	1	AS-95

Figure 7.4 A lubrication map.

- Are there any damaged pipes or tubes? Any leakage from valves and fittings?
- Is there any clogging in reversing, multiway valves or any other relevant parts?
- Is there an adequate amount of lubricants at lubricating surfaces?
- Is there any overheating at pillow blocks, transmission gears, or any other essential parts?

7.4.7 Set tentative lubricating standards

Thus, operators decide on types of lubricants, lubricating and inspecting methods, and tools and intervals for each lubricating point. They then write their results into tentative lubricating standards with maintenance personnel's assistance in accordance with lubricating sequences and/or routes.

Lubricating intervals are initially decided by either guesswork founded on experience or vendors' information, if no records from the past are available. Generally speaking, vendors tend to recommend more frequent lubricating than is essentially needed. It is important to extend intervals as much as possible by carefully observing the operating conditions of lubricating surfaces during routine production.

7.4.8 Remedy defective areas and difficult lubricating areas

When operators conduct a trial lubrication and inspection according to tentative lubricating standards, they discover many and various kinds of defective parts along with difficult lubricating areas in their equipment, as defined in Tables 7.3 and 7.4.

For example:

- Neither platforms nor handrails are provided at lubricating points.
- Lubricating points are inaccessible, too low, or too high.
- Lubricating points are overly dispersed. Longer lubrication time is needed.
- Pipes or fittings from lubricating points to lubricating surfaces are damaged.
- Branch pipes have different lengths. Some lubricating surfaces have an excess supply, others have a shortage.
- Lubricants drip from lubricating surfaces and become new sources of contamination.

Locations of these deviations must be labeled with identification tags and written onto the defective area list or the difficult work area list, following the same procedure described in Chap. 5. Figure 7.5 illustrates an example of defective areas discovered during an overall lubrication inspection.

In addition to the above activities, operators carefully monitor actual conditions of lubricating surfaces. In connection with these observations,

TABLE 7.3 Lubrication Related Defective Areas

- Damage, breakage, clogging or contamination in lubricating systems such as reservoirs, tubes, pipes, valves and other fittings
- Clogging or contamination at lubricating points or surfaces
- Excess, shortage or leakage of lubricants
- Oil oxidation, or contamination by dust, dirt, water or any other foreign materials

TABLE 7.4 Difficult Lubricating Areas

Operators feel difficulty or unhandiness when they:
▪ Supply lubricants at lubricating points
▪ Inspect quantity of lubricants
▪ Inspect conditions of lubricating surfaces

lubricating intervals are extended or shortened until the optimal intervals are determined. Similar to Step 2, remedial actions against difficult lubricating areas are taken after restoration of deteriorated and defective parts. Typical examples of these actions are shown in Fig. 7.6.

7.4.9 Set cleaning and lubricating standards

According to the restorative or remedial actions accumulated in the past steps, operators compile tentative cleaning and lubricating standards into a single standard, if the time targets designated by managers are successfully achieved. Table 7.5 shows a typical example of cleaning and lubricating standards. These are still tentative.

In Step 5, these standards and the detailed inspection standards set in each category of overall inspection are compiled into routine maintenance standards. They follow from the comparison of the two inspecting standards prepared by autonomous maintenance and fulltime maintenance personnel.

This compilation of standards is not always the final goal of TPM activities. Operators must continue to make efforts to improve work methods and equipment in order to finish necessary tasks within given time targets, or even before.

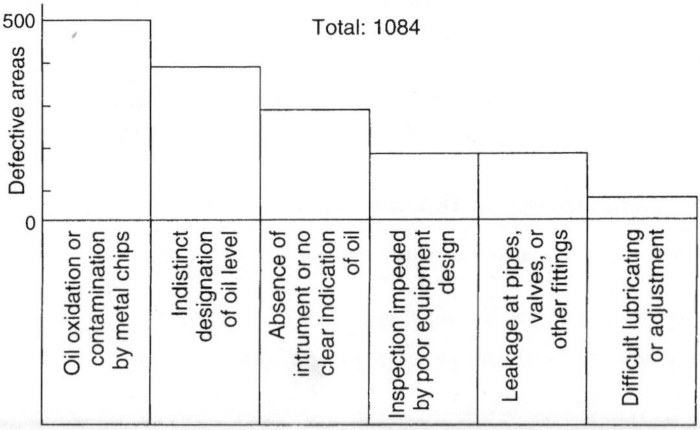

Figure 7.5 Lubrication-related defective areas discovered during an overall inspection.

7.4.10 Thoroughly implement a color lubrication control system

In order to finalize the series of activities described above, operators check once more whether a color lubrication control system is thor-

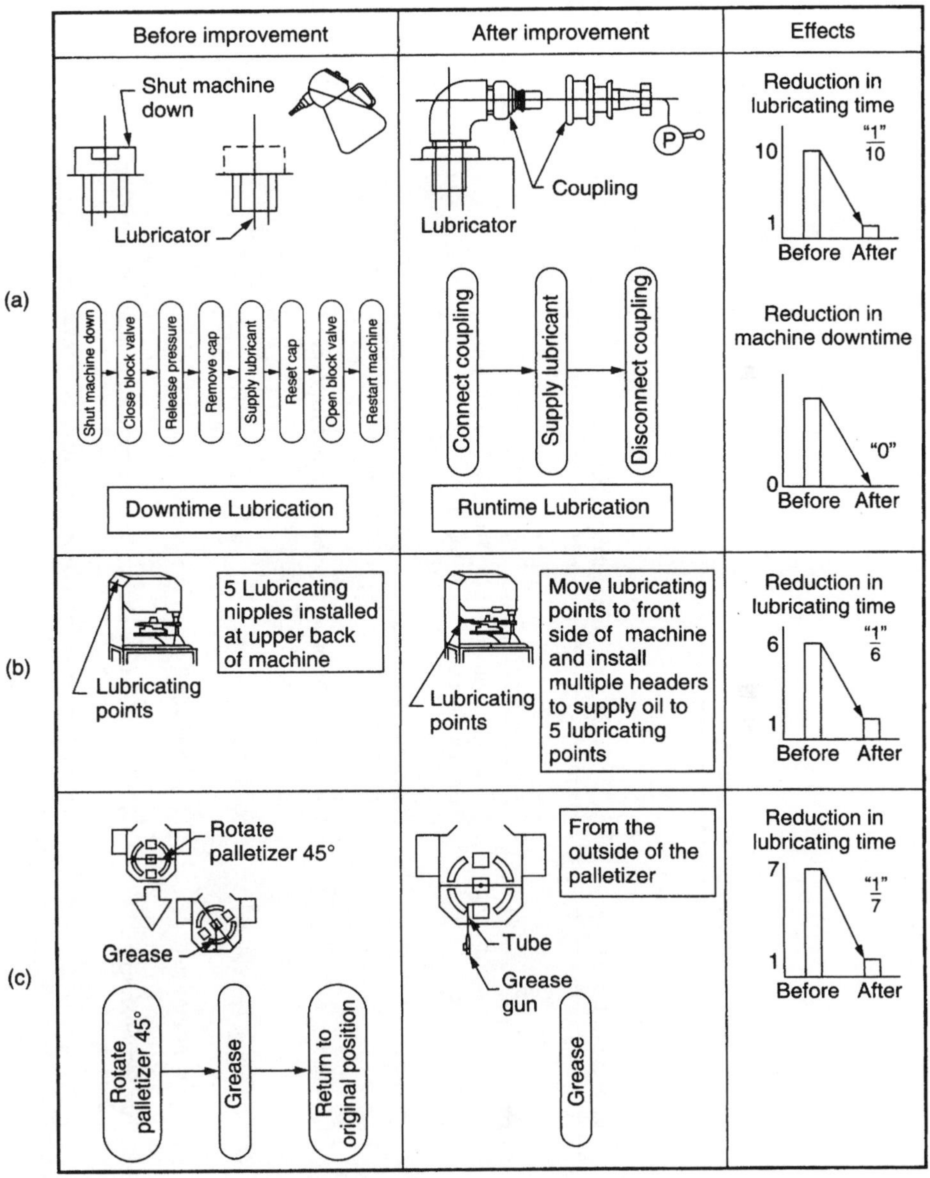

Figure 7.6 (*a*) Runtime lubrication; (*b*) Installing multiple headers; (*c*) Remote greasing.

TABLE 7.5 Cleaning/Lubricating Standards

Equipment: PU-449		Cleaning/lubricating standards		Prepared by: Checked by: Approved by:	Date issued: Date expired: Process:			
					Interval			
	Location	Cleaning standards	Cleaning methods		Minutes	Day	Week	Month
1	Main body and surroundings	No rubber chips or foreign materials	Wipe clean with damp rags		15.		○	
2	Pedestal	No grease or any other contaminants	Remove with scraper and sweep up		5.	○		
3	Peephole	Clear indication of oil	Wipe clean with damp rags		3.		○	
4	Oil pump and valves	No oil leakage or dirt	ditto		10.		○	
5	Underground pit	No oil leakage	ditto		30.			○
	Inspection during cleaning	1) No bolt looseness at ring joints for centralized lubrication system?						○
		2) No bolt looseness or oil leakage at distribution valves?					○	
		3) No bolt looseness at guide plates?						○

TABLE 7.5 *(Continued)*

	Location	Lubrication standards	Type of lubricant	Lubrication methods	Minutes	Interval Day	Week	Month
6	Air lubricator	Within ranges as indicated	Blue-32	Hand oiler	10.		○	
7	Worn gear	ditto	Green-150	Oil container	5.			○
8	Main gear	Sufficient oil film	White-C-2	Scraper	5.			○
9	Oil pump	Within ranges as indicated	Green-68	Oil container	3.	○		
	Inspection during lubricating	1) No play in safety cover for main gear?						○
		2) No anchor bolt looseness for grease supply pump?						○
		3) Adequate oil supply from air lubricator?					○	

oughly implemented to make lubrication and inspection easier and free from mistakes, as follows:

- According to color control rules, paint oil containers, lubricators, and pressure guns with colors assigned to each type of lubricant by indicating viscosity in a suitable position.
- Apply lubrication labels at every lubricating point.
- Mark adequate lubricant ranges at air lubricators and oil level gauges.
- Apply heat-sensitive tape at pillow blocks, speed reduction gears, change gears, or any other critical parts so as to be able to check overheating.

7.5 Case Study

As the overall lubrication inspection proceeded in a machinery manufacturing plant, a large number of defective areas were discovered, as shown in Fig. 7.7. Analysis of these findings disclosed that most of the defects were mainly caused by oil leakage. This leakage produced a situation wherein 436 L of hydraulic oil, on average, were expended on a monthly basis. Another detailed survey conducted in the same plant revealed that most leakages stemmed from fittings, solenoid valves, and sliding parts, as illustrated in Fig. 7.8.

In response, the operators set a goal to reduce oil consumption to less than one third of the original, as illustrated in Fig. 7.9. With the assistance of maintenance personnel, operators attempted the thorough restoration of problematic parts by learning the mechanisms of fittings, valves, pumps, and solenoid valves, and the application of packings and sealants.

As the result of these operators' efforts, the monthly consumption of oil was reduced, as illustrated in Fig. 7.9, and their goal was achieved. In addition, the elimination of oil leakage resulted in the removal of a

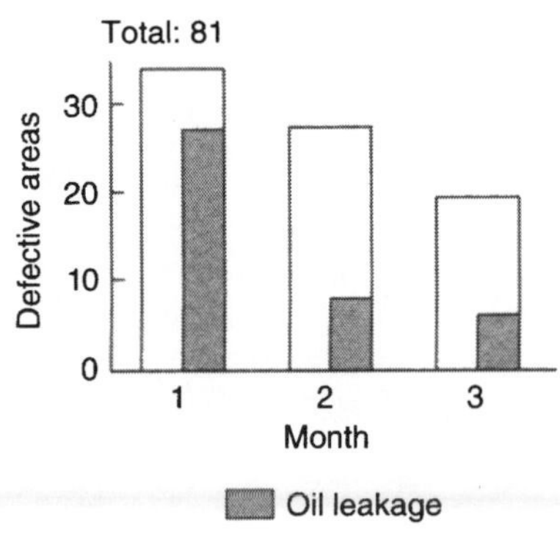

Figure 7.7 Defective areas discovered during an overall lubrication inspection.

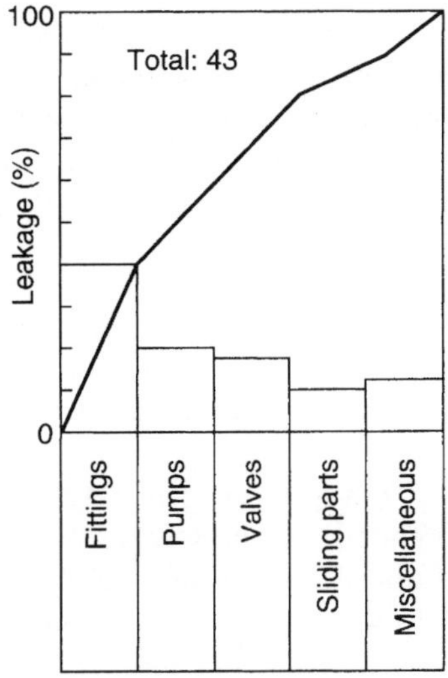

Figure 7.8 A pareto chart of oil leakage.

1.7 m × 1.7 m oil pan from underneath the reservoirs. Operators thereby cleaned more easily the oil supply units and their surroundings within the shortest time frame.

7.6 The Keypoints of an Autonomous Maintenance Audit

This final audit completes the first stage of the autonomous maintenance program aimed at establishing basic equipment conditions. What the auditors must primarily confirm in this audit is how deeply operators have come to understand basic TPM concepts, such as "cleaning is inspection," faithful observation of usage conditions of equipment, and the importance of preventing deterioration.

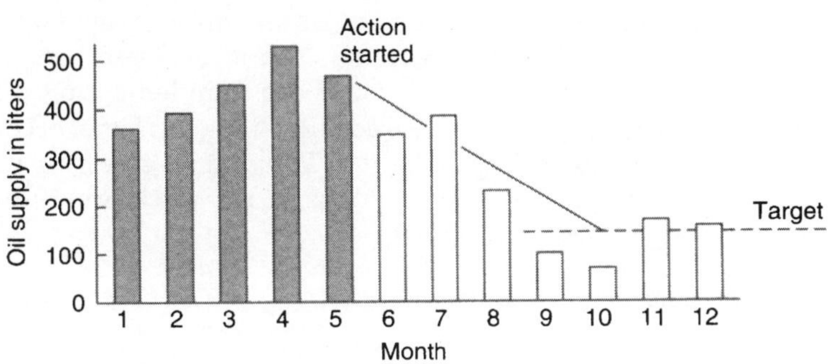

Figure 7.9 A reduction in oil supply.

Also determined is the extent to which each operator begins to secure those skills and habits needed to observe and analyze logically any phenomena in connection with quality defects and breakdowns. If some operators did not make enough progress, suitable supplemental education must be undertaken by each of these operators in consultation with their group leader.

During the on-site audit and subsequent meeting, managers must stress the importance of following the rules established by operators themselves. At the same time, operators should recognize the importance of their own participation in plant operations and the role anticipated of each one of them.

Although operators now are primarily educated about and experienced in the basis of overall inspection in terms of lubrication, they must learn much more about equipment in Step 4. There might, however, be several operators who do not like to learn. In order to get them to participate in TPM activity, it is important for them to understand how each one of them is indispensable to production, as well as to autonomous maintenance activity.

In an on-site audit, managers must make sure that all operators involved in a PM group actually clean and lubricate equipment within the specified time target and in accordance with the standards set by themselves. Table 7.6 shows an example of audit sheets.

7.7 Review the First Stage of the Autonomous Maintenance Program

7.7.1 Establish basic equipment conditions

The first stage of autonomous maintenance is programmed to establish basic equipment conditions. All efforts are concentrated on cleaning and lubricating in Step 3. It is a requisite to understand the overall process of this first stage and the relationship among these three steps by looking at Fig. 7.10.

During the thorough cleaning accomplished in Step 1, operators discover many defective areas in equipment. As cleanliness increases, it is easier to find other problematic locations. Moreover, operators' visual powers are also reinforced. As a result of these complementary effects, the number of defective areas discovered will change in time, according to the curve delineated in Fig. 7.10 (*a*). This curve gives an understanding of why the quick response to operators' work orders by the maintenance department is essential.

7.7.2 Take remedial actions and set standards

At the beginning of Step 2, operators set tentative cleaning standards to maintain the cleanliness audited at the end of Step 1. The time

TABLE 7.6

Step 3	Autonomous Maintenance Audit Sheets	Sheet 1 of 3			
No.	Audit points	Results			
1.	**Step 2 conditions**				
1.1	Is cleanliness attained in Step 2 well maintained? Cleaning standards definitely observed?				
1.2	Are residual issues left over from Step 2 resolved?				
2.	**Group activity** (General)				
2.1	Are aims of Step 3 understood adequately?				
2.2	Is activity plan made in advance? Well executed?				
2.3	Are managers' models well understood?				
2.4	Is TPM activity board adequately utilized?				
2.5	Are safety matters carefully respected?				
2.6	Are TPM activity hours and frequency adequate?				
2.7	Is more efficient way of TPM activity pursued?				
2.8	Are used spare parts and consumables recorded?				
2.9	Is meeting after on-site activity definitely held? Reports submitted?				
2.10	Is activity participated in by all members? No indication of dropout?				
2.11	Are all members cooperating equally? Not led by particular member?				
2.12	Are noteworthy ideas introduced actively to other PM groups?				
2.13	Is cooperation with full-time maintenance satisfactory?				
3.	**Education** (Lubrication)				
3.1	Is education for group leaders and operators conducted? More than 80% of subject matter comprehended?				
3.2	Are lubrication-related devices, parts type, name, structure and function understood?				
3.3	Are phenomena, causes, criteria, evaluation and remedies of lubrication-related equipment defects recognized? Can everyone actually inspect equipment?				
3.4	Does everyone know matters in terms of equipment for which they are responsible as listed here:				

TABLE 7.6 *(Continued)*

Step 3	Autonomous Maintenance Audit Sheets	Sheet 2 of 3			
(1)	Lubricating points and surfaces.				
(2)	Type of lubricants to be applied.				
(3)	Inspection and lubrication intervals.				
(4)	Check points.				
4.	**Overall inspection** (Lubrication)				
4.1	Are all lubricating points and surfaces definitely located?				
4.2	Adequate lubricants at lubricating surfaces?				
4.3	No wear, overheating, abnormal noise or odor at lubricating surfaces?				
4.4	No contaminated, leaked, damaged or clogged grease cups, nipples or any other lubricating points?				
4.5	No pipes, valves and fittings in remote or centralized systems kept in above contitions?				
4.6	Are lubricants not contaminated? Not deteriorated?				
5.	**Tentative lubricating standards**				
5.1	Are lubricating time targets, work allocation and areas clearly specified by manager? Are these prescriptions well understood?				
5.2	Are tentative lubricating standards set immediately after overall inspection?				
5.3	Are tentative lubricating standards revised after every remedial action?				
5.4	Is lubricating work to be done during up/downtime of equipment clearly distinguished? Well understood?				
5.5	Can everyone inspect and lubricate equipment in accordance with tentative lubrication standards? Time targets achieved? If not, are adequate plans and schedules prepared?				
6.	**Cleaning/lubricating standards**				
6.1	Are cleaning/lubricating standards comprised of systematic combinations of tentative cleaning standards set in Step 2 and lubricating standards set in Step 3?				
6.2	Can everyone locate any defects relating to basic equipment conditions?				

TABLE 7.6 *(Continued)*

Step 3	Autonomous Maintenance Audit Sheets	Sheet 3 of 3			
6.3	Can everyone clean and lubricate in accordance with standards? Time target achieved? If not, are adequate plans and schedules prepared?				
7.	**Equipment** (Main body and surroundings)				
7.1	Are tools and jigs stored at designated locations? No shortage or damage?				
7.2	No unnecessary pieces of equipment?				
7.3	No unnecessary materials in process? No dropped parts on floor?				
7.4	Can quality, defective products or scraps be clearly distinguished?				
7.5	Do status display and warning lamps function properly?				
7.6	Do safety devices function properly?				
8.	**Visual control**				
8.1	Are color lubrication control and visual controls such as lubrication labels, instructions for oil levels or adequate ranges of instruments thoroughly applied?				
8.2	Are new visual controls devised? Noteworthy ideas actively revealed to other PM groups?				
9.	**Short remedial program**				
9.1	Is subject selected from the six big losses?				
9.2	Are problems clearly identified? Targets pinpointed?				
9.3	No easy countermeasures taken?				
9.4	Are cost and effects of program reviewed? Actual figures recorded?				
9.5	Are preventive measures against recurrence of problems provided?				
10.	**Residual issues**				
10.1	No residual questions, equipment defects? Are secure plans and schedules for corrective actions prepared?				
10.2	No residual remedies against difficult lubricating areas? If so, do reasonable explanations exist? Are plans and schedules to solve these issues clear?				

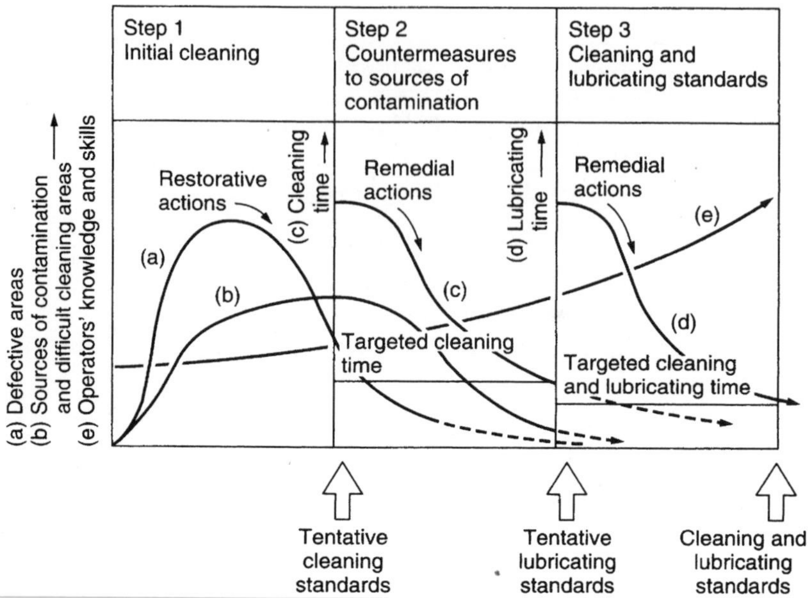

Figure 7.10 Establishing basic equipment conditions.

spent for cleaning, however, is much longer than expected. As these restorative and remedial actions proceed, the time required for cleaning gradually reduces, as illustrated by the curve shown in Fig. 7.10 (*c*). At the end of Step 2, tentative cleaning standards are revised in response to the results of prior activities.

At the end of Step 3, which marks the end of the first stage of the autonomous maintenance program, cleaning and lubrication standards are set to maintain established basic equipment conditions. The time required for both cleaning and lubricating must be shorter than the cleaning time alone attained in Step 2. At this present stage of autonomous maintenance, by way of the repetition of the CAPD cycle, operators gradually achieve the optimal conditions of equipment. Beyond this point, frontline personnel continually aim at a higher level in the spiral, as shown in Fig. 4.3.

Chapter

8

Step 4: Overall Inspection

8.1 Aims from the Equipment Perspective

Prior to the commencement of Step 4, inspection categories, such as fastener, electrical, power transmission, hydraulics, and pneumatics, must be selected. After learning the structure and function of equipment along with inspection methods and criteria for measuring deterioration in each of these categories, operators inspect equipment and remedy defective areas discovered by themselves. When some such restorative or remedial procedures are too difficult for operators, adequate assistance by maintenance personnel is requested.

Operators place appropriate signs or tags in suitable locations on their equipment, for example, names plates, code numbers, or any necessary operating and inspecting instructions on major parts, subassemblies, machines, and processes. Of course, they modify difficult inspection areas, and implement not only established, but also new, visual controls of their own invention.

At the end of each inspection category, they set tentative inspecting standards in order to maintain the equipment conditions achieved by such efforts. This step strives for the improvement and assured maintenance of equipment reliability.

8.2 Aims from the Human Perspective

In terms of each inspection category, operators are educated in requisite inspection and basic maintenance skills so as to be able to detect deterioration and abnormalities in equipment. Through these educational strategies and the subsequent overall inspection, operators understand the importance of making a plan and reviewing the results based on accurate figures reached by collecting and analyzing data in association with actual operating conditions of equipment.

Thanks to a series of these instructional and practical experiences, basic preparation is provided for training knowledgeable operators who will be capable of autonomous supervision involving the repetition of the CAPD cycle in connection with any appropriate number of inspection categories at short intervals.

8.3 The Necessity of Overall Inspection

8.3.1 The reality of inspection

It is very doubtful that traditional inspections conducted by operators in many factories are actually useful for the operation and maintenance of equipment. On closer observation, it is not unusual to find that some operators write inspection results for many days onto check sheets at one time, or place their signatures there without having conducted the specified inspection.

The reasons why operators do not follow the rules is described in Chap. 4. For quite similar reasons, operators' inspections were not effective. Operators could not properly inspect equipment in spite of their possible desire to possibly do so, because the necessary feasibility, motivation, skill, and circumstances were lacking.

These situations are found in many companies where the maintenance staff formulates the check sheets and then attempts to force the operators to perform the actual inspection of equipment. Too much reliance on this staff under such conditions is bound to result in unrealistic and ineffective inspection. Needless to say, an essential reason for operators' noncompliance resides again in the discrepancy between those who set the rules and those who must follow them.

In addition, due to the warranty on machines against failure in performance, vendors feel obliged to demand excessive lubrication and inspection by the user. Besides, the staff in charge tends to be overly cautious in these regards. In the long run, the documents for inspection prepared by them include too many checkpoints without careful consideration of the operators' work load and skills. Nevertheless, this same staff assumes that their task is completed once they give the check sheets to the production department and leave routine inspection up to the operators.

Such situations obviously exist in a significant number of plants operated under poor maintenance conditions. Maintenance personnel who are always pressed to deal with sudden and frequent breakdowns have no time to perform routine inspections. They are called in only when breakdowns occur, and they barely have enough time to replace broken parts without having to check for deterioration in related parts. Furthermore, ordinary operators can never perform effective inspections because of their backgrounds. As a result, it is a reality in many companies that no one responsibly conducts reliable routine inspection.

8.3.2 No motivation is promoted

Inspecting standards prepared by managers or engineers and forced upon operators tend to reflect a lack of careful consideration for the people who must follow them, as shown here.

- What must be inspected?
- Is the allocation of inspecting tasks between production and maintenance personnel adequate?
- Is it feasible for operators to conduct inspection simultaneously with cleaning and lubricating?
- How much time is needed to inspect?
- How long are inspecting intervals extended?
- Are there any ideas to make inspection unnecessary or easier?
- What kinds of skills do operators need to be able to inspect?

In the absence of the above considerations, most standards become impractical and ineffective. In addition, operators usually are not taught about the necessity of inspection and the negative impact that insufficient inspection has on equipment and quality. By neglecting such essential considerations, management demands that the operators conduct proper and faithful inspection. It must reasonably follow that operators do not generate any motivation under these conditions.

8.3.3 No skill is provided

The largest obstacle to ensuring effective inspection of equipment is a lack of skill on the part of those assigned to the task. Inspection to be done by operators is mainly a matter of visual inspection, plus some monitoring of mechanical conditions using simple instruments. Because most of the criteria for operators' inspection depends on the five senses, which are difficult to quantify, the results of these sensory inspections are more difficult to evaluate than are conditions measured by using instruments.

To master proper visual inspection and its evaluation, considerable practice is required. Operators never properly inspect equipment if they are given only inspection check sheets and standards. In other words, there are certain limits to describing in written documents or teaching with audiovisual materials subjects such as normal and abnormal ranges of abrasion, play, noise, temperature, vibration, odor, and so on. To compound matters, maintenance personnel and vendors' specialists occasionally present different answers to questions that arise in reference to the same visual inspection.

What operators need to do, in these regards, is discover abnormalities in equipment during their routine operating, cleaning, and lubricating tasks. An inspection might be based on a kind of intuitive or sixth sense, rather than a structured inspection making reference to check sheets. Frontline managers, therefore, must understand that check sheets are only one of the tools to assure regular and patient inspection of designated areas within a limited time.

From this point of view, it becomes essential, by providing suitable education and practice, to train knowledgeable operators who have a critical, intuitive sense based not on random guessing, but on a technical foundation. After acquiring the necessary skill through such training, operators prepare their own check sheets. Of course, only then is proper inspection achieved by these operators who are capable of establishing rules for routine inspection.

8.3.4 No circumstances are provided

No matter how highly motivated and skilled operators are, they are not psychologically disposed to inspect equipment if they are pressed continuously by their responses to minor stoppages or reworking of off-specification products. Some managers focus only on the quantity of products and, at the same time, surprisingly reduce the time allowed for inspection without a logical basis. This attitude ironically causes an increase in minor stoppages and breakdowns along with reduction in output.

Many managers lack the courage to dare to interrupt production for inspection. Providing the circumstances conducive to inspection is the managers' role. By not making such provisions, they neglect this obligation and compound the problem by making unrealistic demands on operators for satisfactory inspection.

8.3.5 An operator's potential

Education for each operator takes place in terms of each inspection category. In spite of the basic knowledge obtained in the previous steps, operators may experience difficulties with remembering, in detail, the names of parts and the meaning of technical terms. PM group leaders must conduct roll-out education in easily understood language. The same kind of careful attention to teaching materials is required of engineers. The educational process becomes too difficult if the level of education is not raised gradually in proportion to an operator's technical progress.

When operators definitely master inspection skills, they fully appreciate the importance of their patient efforts, made from the initiation of the autonomous maintenance program for the purposes of establishing the basic equipment conditions and preventing deterioration. As a result, they commit themselves to the important role of routine inspection within the context of actual equipment conditions. By this approach, it is feasible to foster the development of knowledgeable operators.

At this time, top management along with factory leadership may still mistakenly underestimate operators' potential. Admittedly, the education of operators must begin with very basic matters at the beginning of TPM activities because inadequate education was provided in the past. Operators' technical skills, however, increase remarkably by means of a series of instructions and the subsequent practice that is repeated throughout Step 4.

Education in autonomous maintenance focuses neither on training a small number of excellent operators nor on eliminating dull operators. It aims, instead, at raising the technical level of all operators, as illustrated in Fig. 8.1. It is, therefore, important to monitor carefully the extent of the operators' understanding and to provide them with more advanced instructions, commensurate with their current progress.

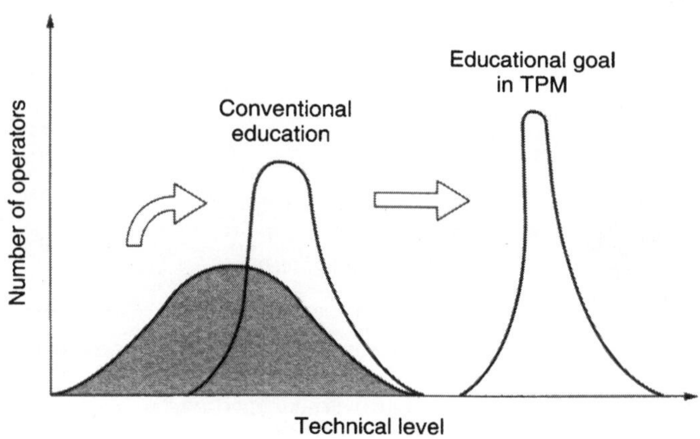

Figure 8.1 Improving operators' technical levels.

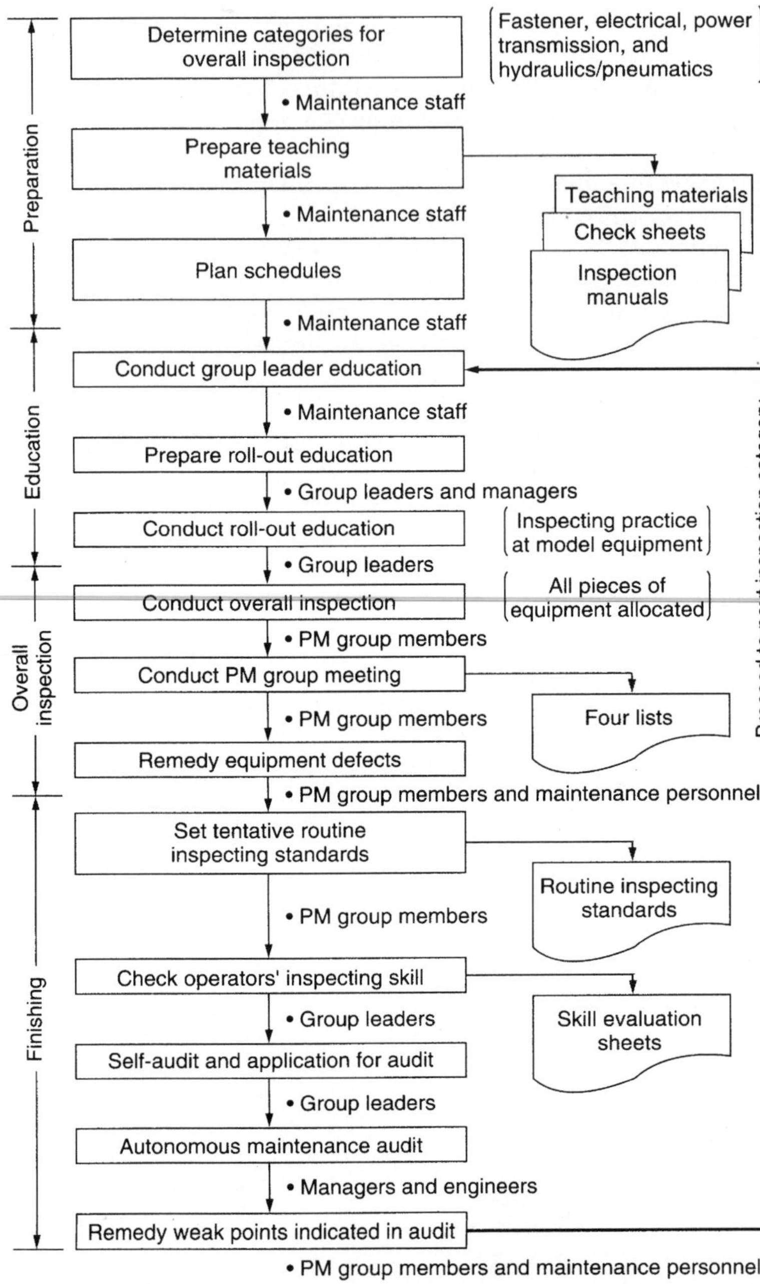

Figure 8.2 The procedure for developing an overall inspection.[1]

8.4 How to Develop an Overall Inspection

8.4.1 Overall inspection procedures and substeps

After education in each inspection category, operators apply newly acquired skills by actually inspecting equipment and then by restoring any defective areas discovered. They reinforce, by practice, the knowledge and skill which they learned. In this way, not only educational goals are reached, but also the complete elimination of equipment defects is accelerated.

As diagrammed in Fig. 8.2, a series of educational and experiential activities, in connection with preventive measures followed by standardization and the autonomous maintenance substep audit, is repeated until all of the specified inspection categories are covered. During this regimen, operators, frontline managers, and engineers are trained.

In most of the companies implementing TPM, a time of at least one year is spent developing this overall inspection activity, which is divided into the substeps listed in Table 8.1. By frequently repeating

TABLE 8.1 Dividing Step 4 into Substeps

Substep	Major activity
1-1	Conduct general education.
1-2	Conduct specific education.
2-1	Conduct a comprehensive test.
2-2	Prepare inspecting check lists.
3-1	Assess inspecting items.
3-2	Conduct an overall inspection.
3-3	Remedy defective areas discovered.
4	Set tentative inspecting standards.
5	Estimate inspecting intervals.
6	Set inspecting time targets.
7	Set improvement targets.
8-1	Identify difficult inspecting areas.
8-2	Remedy difficult inspecting areas.
9	Review inspecting standards.
10	Allocate routine inspecting tasks.
11	Check operators' inspecting skill.
12	Develop a short remedial program.
13	Conduct an autonomous maintenance audit.

the CAPD cycle in a relatively short period of time, all employees certainly become accustomed to autonomous supervision.

8.4.2 Preparing inspection education

1. *Determine a category for overall inspection.*

   The maintenance department determines the category for overall inspection and the basic policy of education. It also makes allowances for requests made of operators on how and what portion of equipment is to be inspected, plus the necessary know-how to achieve the proper operation, setup and adjustment, and routine maintenance of equipment. A more detailed analysis of these matters must be conducted for each plant in accordance with the characteristics of its equipment, configuration of processes and work methods, deterioration of equipment to be inspected, occurrence of quality defects and breakdowns, and so on.

   In addition to the lubrication inspection carried out in Step 2, it is recommended that at least four or more categories of inspection are prescribed in which the CAPD cycle is repeated a minimum of five times. The potential categories of choice include fastener, power transmission, electrical, instrumental, hydraulics, pneumatics, water, steam, and any others which are widely applied throughout plant. (See Fig. 4.4.)

   The selection of an actual inspection category differs with the current operating conditions of major equipment and the policy for autonomous maintenance activity. This book is, however, written on the assumption that fastener, electrical, power transmission, and hydraulics/pneumatics are selected as the overall inspection categories. Basic items of inspection are summarized in the audit sheets attached at the end of this chapter. Further details must be specified in the overall inspection check sheets and manuals prepared by the maintenance department.

2. *Prepare teaching materials.*

   The most fundamental teaching materials to be prepared by the maintenance department are the check sheets and manuals for overall inspection. In terms of each category, maintenance engineers provide the details of inspection to be addressed by operators, and then generate check sheets, as illustrated in Table 8.2, in response to the assumption that operators will perform visual inspection and some other inspection using relatively simple measuring instruments.

   Maintenance engineers then examine the details of the educational program in order to make certain that operators are able to inspect equipment. After these details are set, they should be clearly

TABLE 8.2 Inspection Check Sheets (Partial)

Inspection check sheets Note: For details, refer to "Electrical Section" of inspection manuals Parts type / Check points	Wiring	Push-button, cut-off switches	Limit switches	Precision snap-acting switches	Proximity switches	Photoelectric switches	Auto switches	Relays	Fuses	Terminals	Thermal relays	Ammeters, voltmeters	Motors
Improper actuation due to misalignment			○	○	○	○	○						
Harmful damage, crack, deformation or melting	○	○	○	○	○	○	○			○			○
Loss or looseness	○	○	○	○	○	○	○	○	○	○	○	○	○
Overheating, abnormal noise or odor									○				○
Adherence of cutting fluid, metal chips, dust or spatters	○	○	○	○	○	○	○	○	○	○	○	○	○
Overstress or chafing caused by machine motion	○												
Shock or uneven force			○	○	○								
Installation in vibrating area	○	○	○	○	○	○	○	○	○	○	○	○	○
Proper guard against collision		○	○	○	○	○	○						

TABLE 8.2 Inspection Check Sheets (Partial) *(Continued)*

Inspection check sheets Note: For details, refer to "Electrical Section" of inspection manuals — Check points / Parts type	Wiring	Push-button, cut-off switches	Limit switches	Precision snap-acting switches	Proximity switches	Photoelectric switches	Auto switches	Relays	Fuses	Terminals	Thermal relays	Ammeters, voltmeters	Motors
Possibility of actuation error		○	○	○	○	○	○						
Unnecessary wiring or devices	○	○	○	○	○	○	○	○					
Adjustment ranges properly displayed											○		
Time-consuming adjustment or inspection			○					○					
Proper return of gauge pointer to home position												○	

described by using the easiest language in the overall inspection manuals. These manuals must specify the basic structure, function, nomenclature, inspection methods, criteria, phenomena and causes of deterioration, and remedies in connection with major equipment, subassemblies, and essential parts to be inspected. Figure 8.3 shows an example of a part of an inspection manual.

These teaching materials alone are not adequate to assure the operators' full understanding of the matters listed above. More concrete and visible items selected from familiar machinery installed in the factory, such as typical parts, subassemblies with cutaway models, simplified drawings, sketches, and audiovisual materials are required.

Good materials, however, are not forthcoming if they are arranged for in a hurry shortly before the educational program for operators' inspection begins. It is no easy matter to prepare these kinds of instructional aids. Suitable items may be left over from the so-called TPM workshop which took place to improve maintenance technicians' skills in the early days of full-time maintenance activity.

It is important to begin the careful preparation and collection of additional materials as early as possible by, for example, keeping cutaway models manufactured by maintenance technicians in workshop practice, and by purchasing or making in-house video tapes.

3. *Plan a schedule.*

Maintenance engineers in charge of this program must simultaneously plan the schedule for education during preparation period of the teaching material. The period of time allocated to an inspection category is determined by the difference between the average level of operators' skill in the beginning and the level reached in the future. Many companies allocate several months to any one category of overall inspection.

The schedule must be set up to allow operators to learn prescribed knowledge and skill within a reasonable, but tight, time frame. Careful consideration must be given to the link between education and subsequent practice, lest some operators forget all they learned long before they actually get to the point of inspecting equipment. On the other hand, consideration must be given to the availability of group leaders, the effective allocation of operators' planned overtime hours, the arrangement of training facilities, restrictions due to production plans, and so on.

Of course, the schedule for education cannot be fixed alone by the engineers in charge. It needs to be sufficiently coordinated with pro-

Category	Hydraulics	Inspection
No.	5 - 3 - 2	

Structure

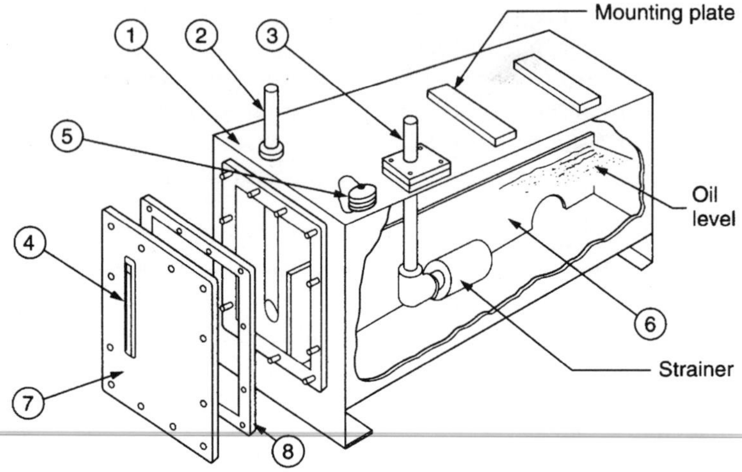

No.	Parts	Services	Inspecting methods
1	Main body	Preserve hydraulic oil	Visual
2	Return pipe	Return oil to reservoir	Visual
3	Pump inlet pipe	Suck oil from reservoir	Visual
④	Oil level gauge	Make oil level visible	Visual
⑤	Air breather		Breathe air plus supply oil
6	Baffle plate	Separate contaminants	Visual
7	Side plate	Removable to clean inside	Visual
⑧	Gasket	Seal connecting portion	Visual

Figure 8.3 An inspection manual.

Manual	Equipment	Reservoir
	Date issued	

Basic function

A reservoir preserves needed hydraulic oil. In addition, it removes contaminants such as dust, dirt, and air, and maintains the oil temperature within a desirable range. Circulation pumps, motors or valves are frequently mounted on it.

Causes	Phenomena	Notes
Contamination by foreign substances	Sludge accumulated on the bottom	
Oil temperature	Oil leakage under higher temperature	
Contamination by water	Rust or corrosion	Oil emulsified
Air intermingled	Oxidation	Clean air
Oil deterioration	Sludge accumulated on the bottom	
Improper installation or loose connections	Oil leakage	
Improper operation	Oil level drop or overflow	Negligence of oil supply

Defective conditions	Inspecting criteria	Corrective actions
Oil leakage or interior reservoir rusted	No looseness, crack, or corrosion	Cleaning tightening or repair
Air bubbles	No looseness or damages	Tightening or repair
Pump cavitation	No looseness or damages	Tightening or repair
Leakage or obscure level of oil	No leakage or sight glass contaminated	Cleaning or repair
No cap or filter breakage	Proper air breather setting	Resetting or replacement
Dropping or rust	No dropping or rust	Cleaning or repair
Oil leakage	No looseness	Tightening
Oil leakage	No deterioration or damage	Replacement

duction department managers. Because this educational program influences budgetary estimates and allocations, the TPM office must actively participate and assist in this planning. This kind of coordinated planning is important, especially in view of the fact that the educational program is a long-term process. Careful planning helps to assure that the program does not collapse along the way. Prior approval by top management might be required. Figure 8.4 is an example of an educational schedule.

Production department managers, in particular, struggle with arranging the assignments of and replacements for PM group leaders while they participate in the educational program. These same group leaders serve as the supervisors or forepersons who are the key personnel in routine production as well as in TPM activities. Furthermore, they generally work on a shift basis. Managers under these circumstances must make the utmost effort to assist group leaders in getting to the classroom. Figure 8.5 gives an example of how group leaders select any class at their convenience from among repetitive courses varying in time of day and day of week.

4. *Execute education.*
In terms of education and practice for all operators, the rollout education method described in Chap. 4 must be applied.

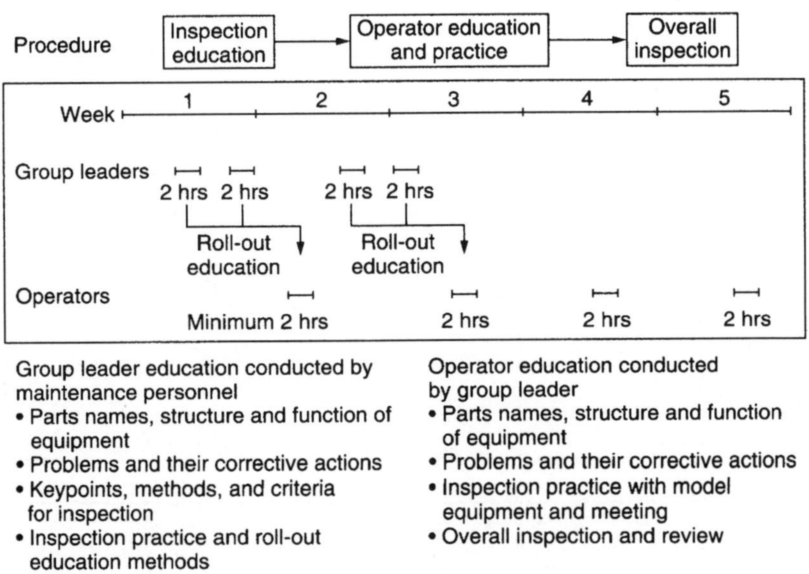

Figure 8.4 An educational schedule for an overall inspection.

Types	Week	Day \ Courses	1	2	3	4	5	6	7	4	8	9	Time per inspection category	Notes
		Time	10:00 ~ 12:00	14:00 ~ 16:00	17:00 ~ 19:00	10:00 ~ 12:00	10:00 ~ 12:00	14:00 ~ 16:00	17:00 ~ 19:00	10:00 ~ 12:00	10:00 ~ 12:00	14:00 ~ 16:00		
Group leader education	1	Mon	EL1	EL1	EL1								6 hours (2 hrs x 3 courses) assembly education	Group leaders are requested to participate in any courses among EL1, EL2, and EL3 at their convenience
		Tue	EL2	EL2	EL2									
		Wed	EL3	EL3	EL3									
		Thu				EL1								
		Fri				EL2								
	2	Mon					EL1	EL1	EL1					
		Tue					EL2	EL2	EL2					
		Wed					EL3	EL3	EL3					
		Thu								EL3				
	3	Mon									EL1	EL1		
		Tue									EL2	EL2		
		Wed									EL3	EL3		
Roll-out education	4	Mon–Fri	←			→							6 hours (2 hrs x 3 courses) roll-out education	In case of three shifts
	5	Mon–Fri					←			→				
	6	Mon–Fri									←	→		
Overall inspection	7	Mon–Fri	←			→							Approximately 4 to 6 hours	
	8	Mon–Fri					←			→				
	9	Mon–Fri									←	→		

Note: This curriculum is planned based on an assumption each course has approximately 20 participants.

Figure 8.5 A curriculum of electrical education for group leaders.

8.4.3 Conducting overall inspection education

1. *Conduct group leader education first.*
The keypoint of this educational program is that PM group leaders must be taught not only the approach to and methods of overall inspection, but also how to perform roll-out education successfully to be able to convey their knowledge to operators afterward. Maintenance engineers must, therefore, enable group leaders to achieve this by using cutaway models, drawings, sketches, and other visual aids.

2. *Prepare roll-out education.*
Group leaders should not simply roll out to their members only what they were taught, but should carefully prepare their own educational program and teaching materials attuned to their equipment and with the assistance of maintenance personnel. To this end, they should try to become aware of and thoroughly understand any obscure issues until they teach operators with confidence.

3. *Conduct roll-out education.*
In roll-out education, lectures are presented by group leaders and pertinent pieces of equipment are inspected by group members. During this practice with inspection, operators learn the structure and function of equipment along with names of parts, methods and criteria of inspection, deterioration, and remedies. Any unintelligible or ambiguous matters found in this course must be written into the question lists. It is important that no residual minor questions are carried over to the next step. In many plants, a brief test is conducted to check on operators' comprehension.

   It might be convenient to conduct the course for roll-out education with two PM groups rather than one. Two group leaders thereby help one another. Moreover, managers and maintenance personnel who participate in the course as advisors reduce their loads.

4. *Make education enjoyable.*
Education in and practice with inspection are developed over a long period. It is, therefore, important to make this experience enjoyable instead of boring in order to attain desirable results. Various ideas in support of this objective were tried. Some strategies include the actual disassembly and reassembly of simple machines, the hands-on examination of problem sites in the factory, contests in the discovery of defective parts, and so on.

8.4.4 Set tentative inspecting standards

Step 4 aims at two objectives: the enhancement of operators' skills and the restoration of deteriorated parts in equipment. After completing the planned education program for their inspection, operators thoroughly inspect equipment category-by-category. As in previous steps, operators apply identification tags at points where component parts are deteriorated or otherwise defective and at difficult inspection areas. Operators determine specific areas of equipment for routine inspection and then set tentative inspecting standards consonant with inspection sequences and/or work routes of operators based on the results of prior overall inspections.

8.4.5 Restore and improve equipment

A great deal of deterioration and defects in equipment is discovered in an overall inspection. These deficiencies are allocated to all operators belonging to a PM group for restoration or improvement. If areas difficult to inspect are encountered, inspecting methods or equipment must be revised until a targeted time is attained.

As in the previous steps, operators remove identification tags and cross out items on the four lists when restorative or remedial actions are successfully completed. The problems that operators are absolutely unable to handle are referred to the maintenance department. By repeating these kinds of activities several times, all operators gradually enhance their inspecting skill.

Most important in this stage of TPM is the response of the maintenance department to the large number of work orders from the production department requesting assistance with over half of the problems discovered in the progression of operators' overall inspection. Maintenance personnel must respond as quickly and effectively as possible to operators' work orders. Otherwise, PM groups lose their enthusiasm and dedication and, eventually, overall inspection fails.

8.5 Case Study

8.5.1 Overall inspection in terms of types of parts

In a plant, a survey of causes classified by the types of parts involved showed that breakdowns occurred in the following frequency: sensors such as limit switches, proximity switches, and light detectors, plus extended actuators, magnet-type relays, air cylinders, and solenoid valves, as illustrated in Fig. 8.6. Because a large number of these parts

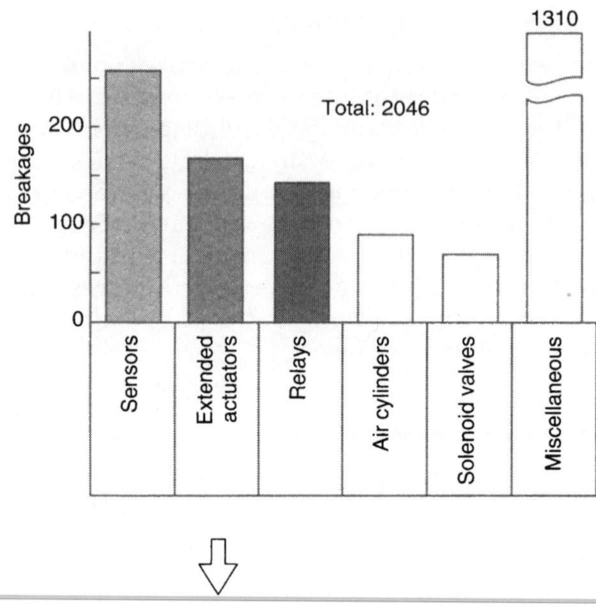

Figure 8.6 A survey of parts' breakage during a three-month period.

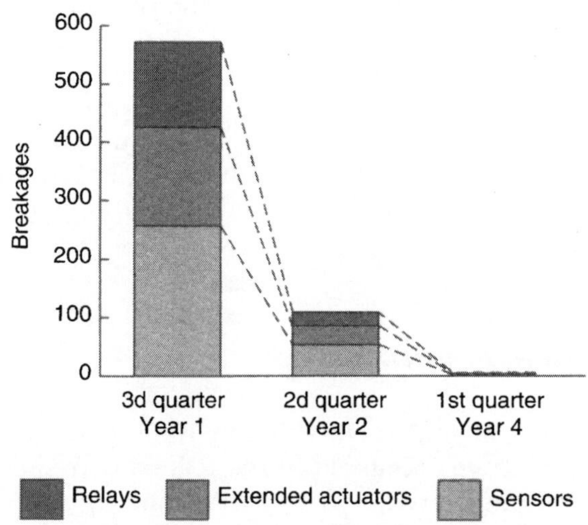

Figure 8.7 A reduction in the breakage of the worst three types of parts illustrated in Fig. 8.6.

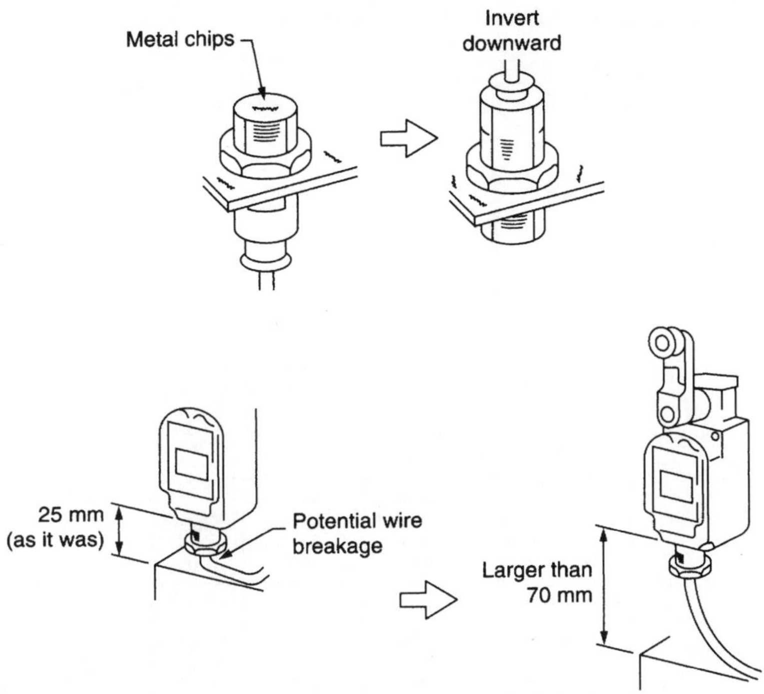

Figure 8.8 Remedial actions taken under the leadership of full-time maintenance.

is located throughout the plant, enormous benefits are expected to result from thorough and proper remedies applied to these defective parts. Out of 23,000 parts inspected during the overall inspection, 5536 defective parts were remedied. Figure 8.7 details the reduction in occurrence of breakdowns caused by the worst three types of parts presented in Fig. 8.6.

In another plant, an overall inspection and subsequent remedies against sensors were conducted during a half-year period under the leadership of full-time maintenance. Maintenance personnel and operators inspected 16,000 parts, and corrected 3132 deteriorated or defective parts. Figure 8.8 shows typical remedial actions.

8.5.2 Overall inspection in terms of categories

In a rubber products manufacturing plant, fastener, electrical, power transmission, and hydraulics/pneumatics were selected as categories for overall inspection. Figures 8.9, 8.10, 8.11, and 8.12 show the results of overall inspection in each category.

• Problems written onto the four lists

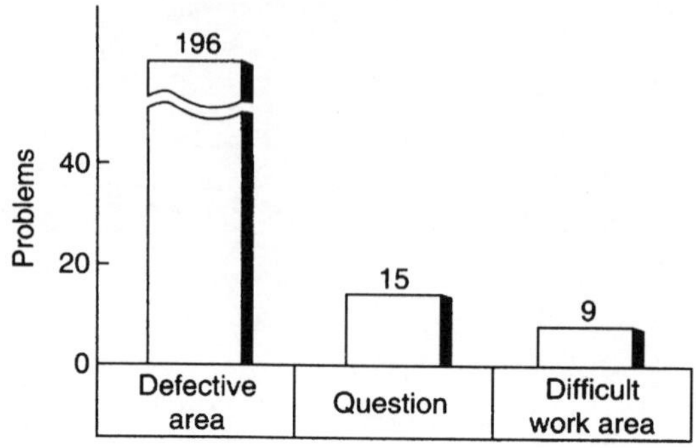

• Defective rates

Equipment	Parts inspected	Defective parts	Rate
Main body	2129	99	4.6%
Auxiliary	232	7	3.0%
Others	824	90	10.9%
Total	3185	196	6.1%

• Details of defective areas

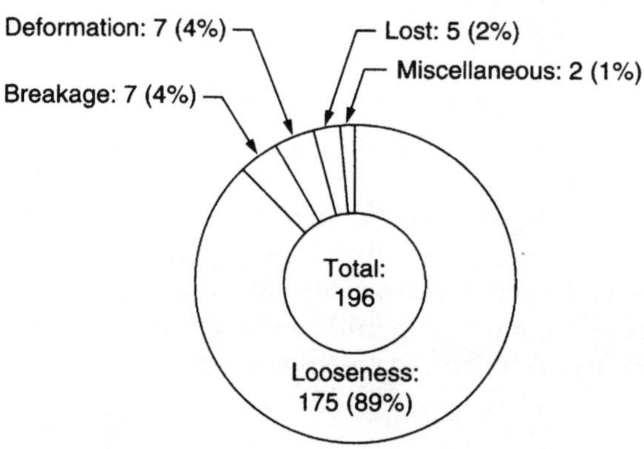

Figure 8.9 The results of an overall fastener inspection.

• Problems written onto the four lists

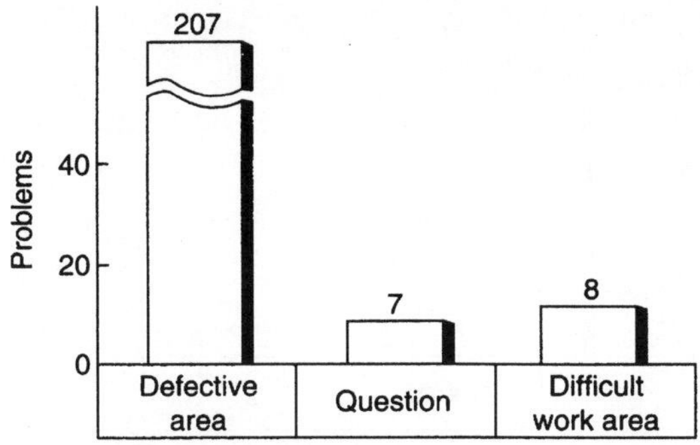

• Defective rates

Equipment	Parts inspected	Defective parts	Rate
Main body	131	55	42.0%
Auxiliary	213	118	55.4%
Others	95	34	35.8%
Total	439	207	47.2%

• Details of defective areas

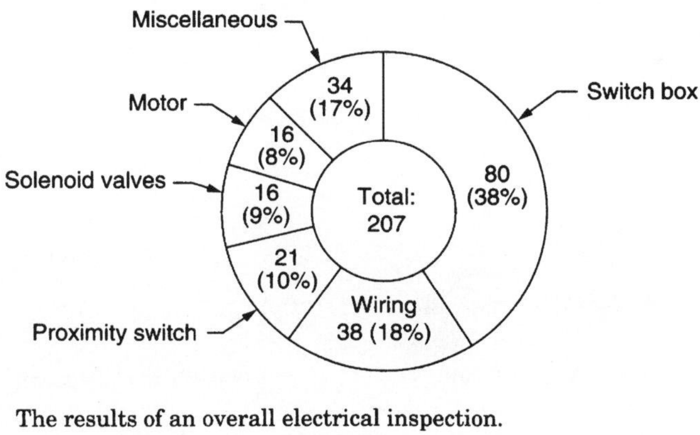

Figure 8.10 The results of an overall electrical inspection.

• Problems written onto the four lists

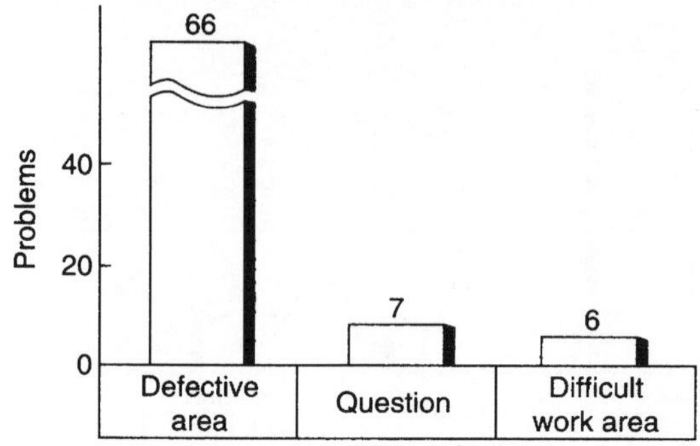

• Defective rates

Equipment	Parts inspected	Defective parts	Rate
Main body	788	39	4.9%
Auxiliary	594	21	3.5%
Others	160	6	3.7%
Total	1542	66	4.3%

• Details of defective areas

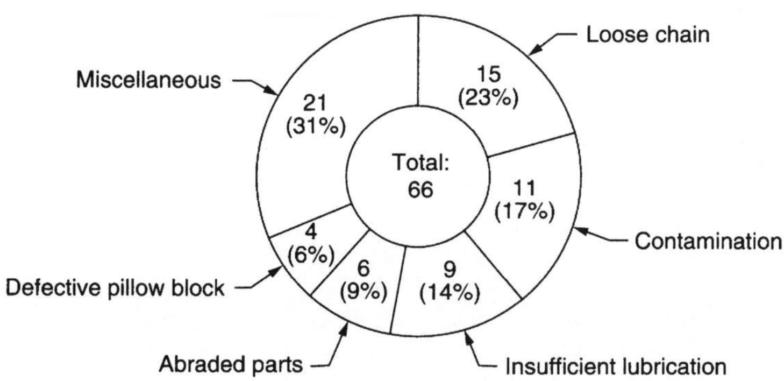

Figure 8.11 The results of an overall power transmission inspection.

- Problems written onto the four lists

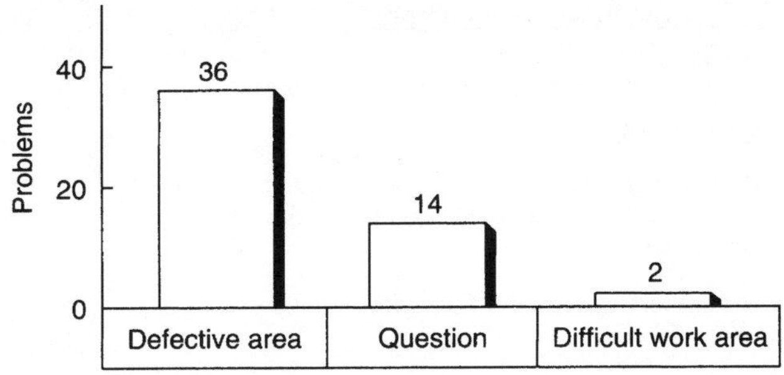

- Defective rates

Equipment		Parts inspected	Defective parts	Rate
1	Main body	205	22	10.7%
2	Auxiliary	110	10	9.0%
3	Others	60	4	6.6%
Total		375	36	9.6%

- Details of defective areas

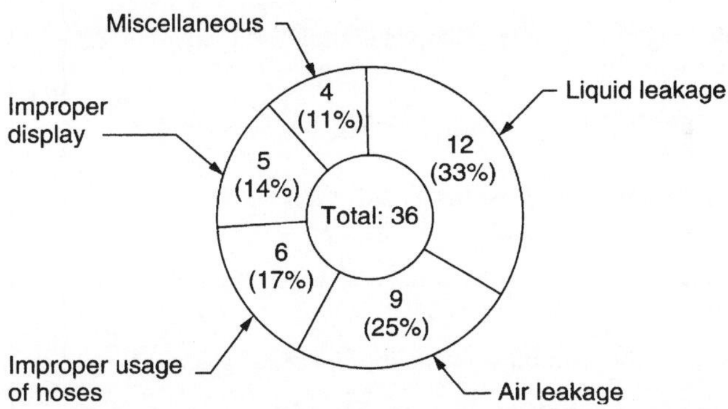

Figure 8.12 The results of an overall hydraulic and pneumatic inspection.

TABLE 8.3

Step 4	Autonomous Maintenance Audit Sheets	Sheet 1 of 5			
No.	Audit points	Results			
1.	**Step 3 conditions**				
1.1	Are the basic equipment conditions attained by Step 3 well maintained? Cleaning/lubricating standards definitely observed?				
1.2	Are residual issues left over from Step 2 or previous substep resolved?				
2.	**Group activity** (General)				
2.1	Are aims of Step 4 understood adequately?				
2.2	Is activity plan made in advance? Well executed?				
2.3	Are managers' models well understood?				
2.4	Is activity board adequately utilized?				
2.5	Are safety matters carefully rspected?				
2.6	Are TPM activity hours and frequency adequate?				
2.7	Is more efficient way of TPM activity pursued?				
2.8	Are used spare parts and consumables recorded?				
2.9	Is meeting after on-site activity definitely held? Reports submitted?				
2.10	Is activity participated in by all members? No indication of dropout?				
2.11	Are all members cooperating equally? Not led by particular member?				
2.12	Are noteworthy ideas introduced actively to other PM groups?				
2.13	Is cooperation with full-time maintenance satisfactory?				
3.	**Inspection education** (By each category)				
3.1	Is education for group leaders and operators conducted? More than 80% of subject matter comprehended?				
3.2	Are parts type, name, structure and function of equipment understoood?				
3.3	Are phenomena, causes, criteria, evaluation and remedies of equipment defects understood? Can everyone actually inspect equipment?				

TABLE 8.3 *(Continued)*

Step 4	Autonomous Maintenance Audit Sheets	Sheet 2 of 5			
3.4	Does everyone know relevant inspecting items and check points in equipment for which they are responsible?				
4.	**Overall inspection** (By each category) • Use attached sheets for each inspection category.				
5.	**Tentative inspecting standards** (By each category)				
5.1	Are inspecting time targets, work allocation and areas clearly specified by manager? Are these prescriptions well understood?				
5.2	Are tentative inspecting standards set immediately after overall inspection?				
5.3	Are inspecting standards revised after every remedial action?				
5.4	Is inspection to be done during up/downtime of equipment clearly distinguished? Well understood?				
5.5	Can everyone conduct routine inspection in accordance with tentative inspecting standards? Can defective parts actually be discovered?				
5.6	Are inspecting intervals estimated?				
5.7	Are time targets achieved? If not, are adequate plans and schedules prepared?				
6.	**Equipment** (Main body and surroundings)				
6.1	Are tools and jigs stored at designated locations? No shortage or damage?				
6.2	No unnecessary pieces of equipment?				
6.3	No unnecessary materials in process? No dropped parts on floor?				
6.4	Can quality/defective products or scraps be clearly distinguished?				
6.5	Do status display and warning lamps function properly?				
6.6	Do safety devices function properly?				
7.	**Visual control**				
7.1	Are visual controls thoroughly applied to facilitate inspecting tasks?				
7.2	Are new visual controls devised? Noteworthy ideas actively revealed to other PM groups?				

TABLE 8.3 *(Continued)*

Step 4	Autonomous Maintenance Audit Sheets	Sheet 3 of 5			
8.	**Short remedial program**				
8.1	Is subject selected from the six big losses?				
8.2	Are problems clearly identified? Targets pinpointed?				
8.3	No easy countermeasures taken?				
8.4	Are cost and effects of program reviewed? Actual figures recorded?				
8.5	Are preventive measures against recurrence of problems provided?				
9.	**Residual issues**				
9.1	No residual questions, equipment defects? Are secure plans and schedules for corrective actions prepared?				
9.2	No residual remedies against difficult inspecting areas? If so, do reasonable explanations exist? Are plans and schedules to solve these issues clear?				

TABLE 8.3 *(Continued)*

Step 4	Autonomous Maintenance Audit Sheets	Sheet 4 of 5			
4.	**Step 4-1 (Overall fastener inspection)**				
4.1	No corroded, loose or bent nuts and bolts? Minimum two threads exposed? No damaged threads?				
4.2	Are lengths of bolts adequate?				
4.3	Are looseness prevention methods such as washers properly used?				
4.4	Are double nuts or match marks used for critical bolts?				
4.5	Are other fastener-related parts inspected?				
4.6	Are all discovered defective parts corrected?				
4.	**Step 4-2 (Overall electrical inspection)**				
4.1	No contamination, scratched, bent or loose connections at cables and conduit pipes? None touching water, steam or moving items? No threat to safety?				
4.2	No disconnections or breakages in ground wires?				
4.3	Do instruments installed on consoles, control panels or distribution boards function properly? No burned out lamps? Are switches easy to use? No possibility of operation mistakes? Are visual controls properly applied?				
4.4	No dirt, dust or any other contamination in/outside consoles, control panels and distribution boards? Are wirings neat? No unnecessary materials left in boxes? Sufficient ventilation?				
4.5	Do motors produce no abnormal heat, noise or odor? No loose setting bolts? Is lubrication satisfactory?				
4.6	No contaminated or damaged limit, proximity switches or any other sensors? No improper actuations?				
4.7	Are all electrical and instrumental parts, devices, wiring, or conduit pipes inspected?				
4.8	Are all discovered defective parts corrected?				
4.	**Step 4-3 (Overall power transmission inspection)**				
4.1	No deteriorated or oil contaminated belts? Are belt tensions proper? No worn or misaligned pulleys?				
4.2	No wear or slack in chains or sprockets? No oily contamination around chains?				
4.3	No damage, misalignment, loose fixtures or play in shaft? No overheating, vibration or noise at bearings?				

TABLE 8.3 *(Continued)*

Step 4	Autonomous Maintenance Audit Sheets	Sheet 5 of 5			
4.4	No noise, vibration or overheating at gears, breaks?				
4.5	Are critical areas visible through safety cover? Are covers adequately modified?				
4.6	Are all other power transmission parts and devices inspected?				
4.7	Are all discovered defective parts corrected?				
4.	**Step 4-4 (Overall hydraulic/pneumatic inspection)**				
4.1	No contamination, damage, leakage or vibration in hydraulic and pneumatic equipment, pipes, hoses, valves, fittings or any other parts?				
4.2	No abnormal noise, vibration or overheating in rotary machinery, motors, solenoid and selector valves, or any other hydraulic and pneumatic parts?				
4.3	Are air filters, regulators and lubricators properly installed? Easy to inspect for oil quantity? Oil levels satisfactory? No foreign particle in filters? Condensed water drained?				
4.4	Are pressure gauges, thermometers, level gauges properly installed? Easy to inspect? Actuating normally?				
4.5	Are tubes, pipes and hoses properly installed?				
4.6	Are all other hydraulic/pneumatic parts and devices inspected?				
4.7	Are all discovered defective parts corrected?				

8.6 The Keypoints of an Autonomous Maintenance Audit

The autonomous maintenance audit is conducted at the end of each substep allocated to an inspection category. An overall inspection is a comprehensive and systematic process for providing education and practice with inspection for operators. The CAPD cycle is frequently repeated over a short period of time in order to enhance operators' autonomous supervision ability. The items for and types of inspection to be compiled on the audit sheets reflect the requests previously made of operators as part of their routine tasks to inspect selected parts of equipment.

Step 5 is the most important stage of the seven-step program in terms of education for operators. Frontline managers who mainly perform audits, however, face a very heavy work load. Audits must be repeated night and day for PM groups working on a shift basis. Careful attention, therefore, needs to be paid to prevent some particular managers from becoming overloaded. Considerable assistance from plant engineering and maintenance department managers might be necessary to alleviate this potential problem.

In the meantime, while overall inspection proceeds, the occurrence of breakdowns is rapidly reduced. Such favorable results encourage not only operators, but also managers and supervisors. Managers must continue to teach operators, as well as themselves, to develop new activities without forgetting the starting point of autonomous maintenance: maintenance of basic equipment conditions, "cleaning is inspection," etc.

It is also important for operators to gradually come to understand the relationship between the current activity focused on Zero Breakdowns and the matter of quality to be addressed in Step 6. Table 8.3 shows a sample of autonomous maintenance audit sheets.

Note

1. Adapted from Soiichi Nakajima, *TPM Development Program* (Productivity Press, 1989), p. 174.

Chapter

9

Step 5: Autonomous Maintenance Standards

9.1 Aims from the Equipment Perspective

Upon entering Step 5, breakdowns of equipment are decreasing dramatically. In a considerable number of processes, Zero Breakdowns are achieved on a monthly basis. Frontline personnel should now find it possible to maintain dependable and consistent operating conditions, especially compared to those attained in previous steps.

If operators still need to continue to clean, lubricate, and inspect equipment manually, in spite of every endeavor to enhance reliability and maintainability, further efforts must be made to invent other visual controls and mistake proofs to make the required tasks easier and free from human error.

In addition, the roles of human and mechanical components are examined for the purpose of improving operability. Step 5 thereby aims at the realization of an orderly shopfloor where any minor deviation from normal or optimal conditions can be detected at a glance.

9.2 Aims from the Human Perspective

The cleaning and lubricating standards set in Step 3 and the tentative inspection standards prepared for each given inspection category in Step 4 are now combined during Step 5. These two standards are combined into unified "autonomous maintenance standards," which prescribe the necessary routine tasks of cleaning, lubricating, and inspecting to be carried out by operators.

Following these standards, operators, by trial, clean, lubricate, and inspect all equipment installed in their process in the same time frame and in three dimensions. By repeating these efforts, they search for and discover the optimal combination of these routine maintenance procedures by thoroughly understanding the equipment and process as a system to be able to set more effective standards. They then consider the duration and intervals to be allocated to these routine tasks, and learn in the process how to obtain and use data on breakdowns and quality defects.

This approach results in the setting of standards which are easy to follow and produce desirable effects. Step 5 thereby aims at the realization of operators who achieve genuine autonomous supervision by setting and following their own rules.

9.3 Finishing the Activities Relating to Equipment

9.3.1 Review residual issues

Step 5 is the final step for coming to grips with matters pertaining to equipment. In Step 6, the subject of autonomous maintenance activities is expanded from the subject of equipment to that of the entire area of processes and is focused on matters of quality.

In general, there are some unresolved problems or unattained goals which were not addressed by remedial actions taken in prior steps. PM groups, therefore, must check these residual issues in terms of defective areas of equipment, sources of contamination, difficult work areas, extension of work intervals, and so on. In the absence of impediments, such as the necessity of continuous operation without shutdown due to a change in a production plan or a delay in delivery of parts by vendors, PM groups must devise a new schedule for completing these residual issues by the end of Step 5.

Managers assist and carefully arrange for each PM group to finish on schedule this final step dedicated to equipment, as specified in the TPM master plan. In doing so, they must take into consideration the progress made and circumstances experienced by each group. Frontline managers and engineers, thereafter, must make an utmost effort to launch activities pertaining to quality together with all PM groups involved in the production department at the beginning of Step 6.

9.3.2 Obtain total knowledge about equipment

In Step 4, operators are instructed in each inspection category in response to machine elements and common subassemblies widely applied throughout a plant. Beyond this perspective, operators should come to realize that while equipment consists of common and specific functions, it also is actuated as an integrated system.

Arriving at a holistic view of equipment as in integrated system is the final phase of education that involves the entire list of activities targeted for equipment in Steps 1 through 5. This same understanding is indispensable to cleaning, lubricating, inspecting, operating, setting, and adjusting equipment, or any other routine tasks, as well as to developing quality assurance activities in the next step. Beyond moving on to Step 6, frontline managers and engineers need to assess carefully each operator's skill in reference to the evaluating procedure described in Chap. 3. These same personnel must then follow up by helping individual operators reach the required level of skill in response to a consultation with their PM group leaders.

9.3.3 Becoming aware of abnormality by the five senses

In many processes, breakdowns may be reduced to one-tenth or one-twentieth of the bench marks as a natural consequence of various remedial actions taken in the previous steps. Some processes may reach the status of Zero Breakdowns on a monthly basis. It is, however, quite difficult to maintain Zero Breakdowns continuously over a one-year period. Once breakdowns reach these lower levels, they cannot be eliminated further despite the greatest feasible efforts made to maintain basic equipment conditions, and to carefully inspect and restore deteriorated areas of equipment.

Because there is a huge number of parts built into equipment installed throughout an entire plant, it is absolutely impossible to achieve complete inspection and precise estimation of parts life to conduct preventive replacement of these parts. It is crucial to detect signs of abnormality in equipment by means of the operator's five senses.

However, no matter how precisely and persistently operators inspect equipment at this stage of development, an extreme reduction of breakdowns is not yet realized. Operators are limited because they are skilled only in the traditional sense and, as such, detect only deterioration occurring on the exterior of equipment. Even highly skilled maintenance personnel are also unable to be aware of defective parts on the interior of equipment by sensory observation and without the benefit of disassembly.

In order to develop a superior sensory ability resulting in a kind of sixth sense, routine training must be continued patiently at the occurrence of every significant breakdown over a long period of time in light of the considerations listed below:

- Were there any indications of breakdowns?
- Did the said breakdowns have specific indications?
- If so, what kinds of indications were detectable?
- Why were the indications not detected prior to the occurrence of the breakdowns?
- By what means are indications be detected?
- What kind of routine training do operators need in these regards?

In view of the above considerations, prior detection of indications linked to breakdowns is encouraged from the early stage of autonomous maintenance activity. In the event that a certain indication of any abnormality is discovered during operation, the operator reports his or her observation to the maintenance department with a warning card. Some plants refer to this warning card as a "yellow card." Figure 9.1 shows a sample of such a warning card.

Maintenance personnel respond by checking equipment with the same operator, and then taking necessary action. If a potential loss is prevented in this way prior to its occurrence, the operator is evaluated highly in accordance with the corporate suggestion system. Actual cases of this kind of activity are introduced later in this chapter.

9.3.4 Thorough implementation of visual controls

Thanks to the repetition of education and the subsequent practice in overall inspection which took place in Step 4, operators' technical knowledge and skill reached new heights. There is, as a result, no reasonable comparison to be made between the levels met at this point with those in the early stages of TPM. By virtue of these higher abilities, operators once more check and evaluate the visual controls implemented in the past. If inadequate, neglected, or unused visual controls are discovered, operators take suitable remedial actions. They must of course continue to search for new visual control methods.

9.4 How to Develop Step 5

9.4.1 Preparatory procedures for autonomous maintenance standards and substeps

In order to finalize all equipment-related activities and establish autonomous maintenance standards, Step 5 is implemented in the following sequence, as diagrammed in Fig. 9.2.

Yellow card	No. 2-16-4	
	Machining - Assembly Plant	
	Prep'd by B. Smith	PM Group Leader S. Jones

Date issued	2/15/89	Line	G assembly line
Time issued	10h:00	Equipment	TM-1048 Performance tester

Production Dept	Machine inspection order Noise from main shaft at No. 5 station --- Rattling in operation
Maintenance Dept	Results of inspection & corrective actions Keyway of mainshaft at 5-ST corroded. Its misalignment with pulley causes noise. Quality defects will occur, if it is neglected. (Urgent spare parts fabrication and replacement is needed) Spare parts arranged and a purchase order faxed by purchasing dept on 2/15. Delivery is scheduled on 2/17. Arrange repair work in the next week-end maintenance on 2/17, 18.

Inspected by	Date inspected	Personnel	Type	Applicable	Approved by
J. Cruz	2/15	15 min	(A) B C	no	M. Holl

Type of abnormality: **Unusual!**
(Abnormal noise), Vibration, Overheating, Abnormal oil reduction, Odor, Abrasion, Rough actuation, Chafing in wiring or hoses, Poor coolant spray, Changing dimensions, Others

Figure 9.1 Reporting signs of abnormality.

1. *Review cleaning and lubricating standards.*
   Based on the experiences obtained by executing the cleaning/lubricating standards set up in Step 3, cleaning tasks and lubrication control must be reviewed as noted here and necessary remedial actions must be taken.

   Cleaning standards

   - Are matters such as areas to be cleaned, cleaning methods, and intervals and criteria for cleaning adequately covered?
   - Do any sources of contamination remained untreated?

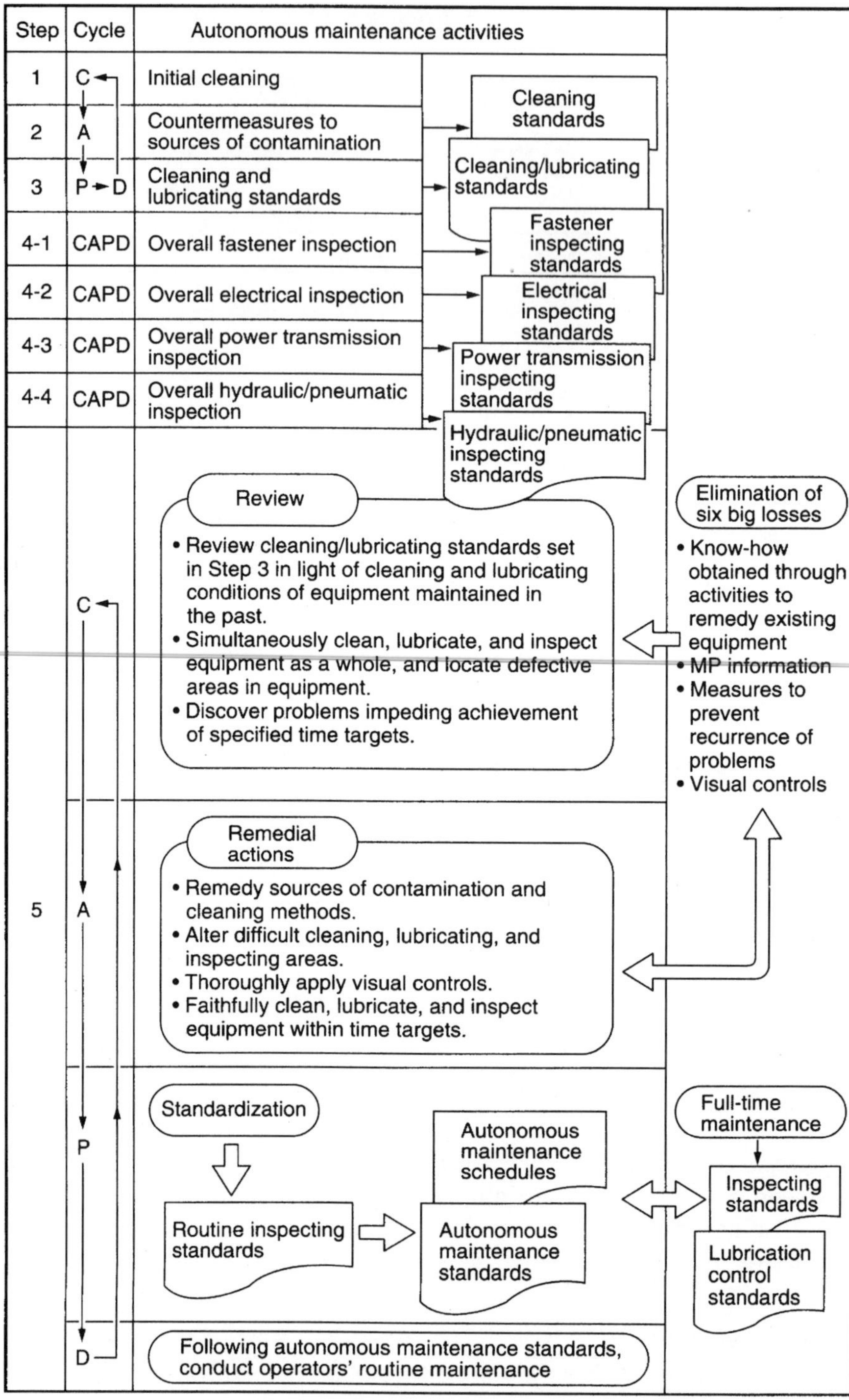

Figure 9.2 The procedure for implementing operators' routine maintenance.

- Are there any sources of contamination newly detected?
- If any problem is discovered, necessary action must be taken.

Lubricating standards

- Are matters such as lubricating points and surfaces, types of lubricants, lubricating methods and intervals, and inspection criteria adequately covered?
- Is the allocating of lubricating tasks between autonomous maintenance and full-time maintenance adequate?
- If any problem is discovered, necessary action must be taken.

2. *Review tentative inspecting standards.*
   Based on the experience gained through the execution of inspecting standards set for each inspection category in Step 4, tentative inspecting standards are reviewed here:
   - Are matters such as inspection location, objective, methods, intervals, and evaluation criteria adequately covered?
3. *Compare with full-time maintenance's inspecting standards.*
   The two separate standards set by the autonomous maintenance and full-time maintenance personnel must be compared to one other to determine a suitable allocation of inspecting tasks. In view of this allocation, operators combine inspecting standards prepared for each category into a single routine inspecting standard for each piece of equipment by reviewing the same procedures just noted.
4. *Set tentative autonomous maintenance standards.*
   The combination of these routine inspecting standards and cleaning and lubricating standards is compiled into tentative autonomous maintenance standards for routine cleaning, lubricating, and inspecting. Operators now must reduce the time spent for this routine maintenance by implementing visual controls and modifying difficult work areas.
5. *Detect the abnormality caused by internal deterioration.*
   Routine inspecting standards are reviewed based on the analysis of warning ("yellow") cards issued in the past, as noted here:
   - Is work allocation between autonomous maintenance and full-time maintenance feasible?
   - Divide past breakdowns and minor stoppages into two groups: one caused by poor maintenance of basic equipment conditions and the other by lack of careful observation, which must be prevented by operators' visual inspection during routine operations.

- Analyze the phenomena of breakdowns and minor stoppages based on the structure and functions of equipment, and take necessary remedial actions.
- Review routine inspecting standards in accordance with the results of remedial actions.
- Make an assessment of operators' skill based on an evaluation system, described in Table 3.1, for the purpose of improving each operator's technical knowledge and skill.

6. *Establish autonomous maintenance standards.*
   - In addition to visual controls, improve equipment and work methods to meet time targets for cleaning, lubricating, and inspecting.
   - Revise and finalize autonomous maintenance standards and schedules.
   - Execute routine operators' maintenance in accordance with autonomous maintenance standards and schedules prepared according to the procedures just mentioned.

The above activities are carried out after dividing them into suitable substeps. Table 9.1 shows a sample of these substeps.

9.4.2 Attain given time targets

The time spent on operators' routine maintenance must be reduced, as illustrated in Fig. 9.3, by continuous endeavors to remedy equipment and improve work methods in order to finish cleaning, lubrication, and inspection within time targets specified by managers. At the same time, the quality of these tasks must be maintained at a higher level than before. Figure 9.4 illustrates a typical example of a reduction in time spent for operators' routine maintenance.

9.5 Routine Inspection by Autonomous Maintenance

9.5.1 Determine the allocation of work between the production and maintenance departments

Prior to the commencement of Step 5, the maintenance department must prepare maintenance standards, as shown in Table 9.2, and schedules from its own perspective in terms of inspection, calibration, replacement of parts, condition monitoring, machine diagnosis, overhaul, etc. In these standards inspecting items which require no disassembling of equipment share exactly the same objectives and methods specified in the inspecting standards prepared by operators in this step.

In addition, many other items may be incorporated into the standards prepared by the maintenance department as follows:

- Inspection attainable by operators once they finish education and practice for inspection in Step 4.
- Special inspections not included in above training, but now appropriately dealt with by operators.
- Items hopefully inspected by operators during routine tasks.
- Reciprocally, items inspected by maintenance personnel requested by operators.
- Maintenance department did not inspect in spite of desire to do so, due to a shortage of work force.

TABLE 9.1

Substep	Major activity
1	Review residual issues left over from Steps 1 through 4.
2	Problems:
2-1	Identify problems in cleaning/lubricating standards set in Step 3 and categorical inspecting standards set in Step 4.
2-2	Plan remedial actions.
2-3	Take remedial actions.
2-4	Evaluate results of actions.
3	Breakdowns and minor stoppages:
3-1	Examine causes.
3-2	Plan remedial actions.
3-3	Take remedial actions.
3-4	Evaluate results of actions.
4-1	Set tentative routine inspecting standards.
4-2	Compare with maintenance standards set by full-time maintenance.
4-3	Set inspecting time targets.
4-4	Set tentative routine inspecting schedules.
4-5	Conduct routine inspection.
5	Review cleaning, lubricating/inspecting standards.
6	Set autonomous maintenance standards and schedules.
7	Develop a short remedial program.
8	Conduct an autonomous maintenance audit.

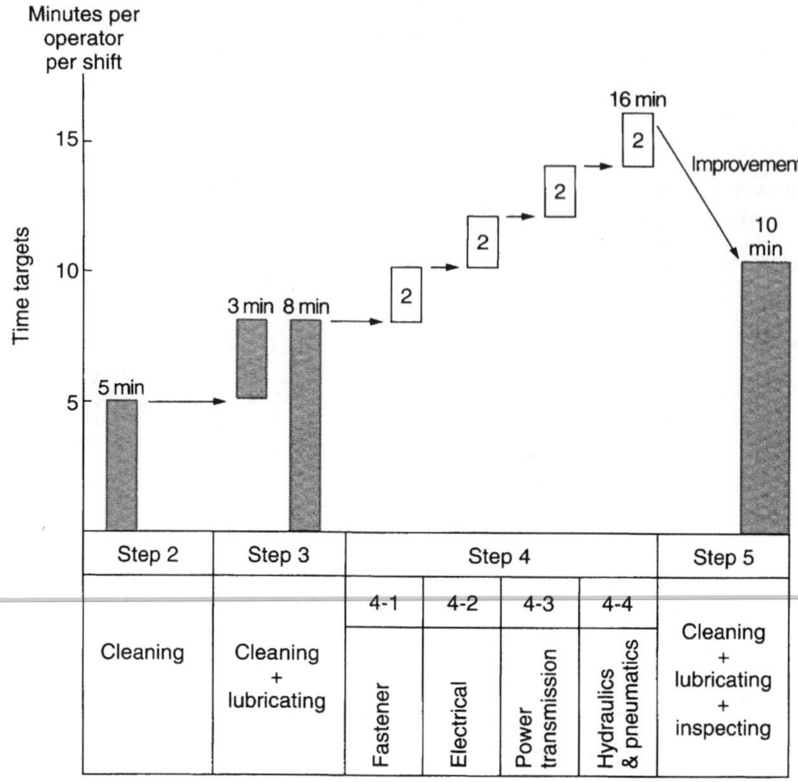

Figure 9.3 Time targets in autonomous maintenance activities.

Surprisingly, the inspecting standards set by operators sometimes are superior to those set by the maintenance department, because no plant undergoes as thorough an inspection as the one carried out in Step 4. Of course, some of them may have weak points. Therefore, the two standards set by autonomous maintenance and full-time maintenance are compared item by item for each piece of equipment.

By thoroughly reviewing these standards for duplicated or omitted inspection items, the allocation of routine tasks between the production and maintenance departments is finally determined. It is important to adjust carefully these two standards, so they supplement each other and blend into a composite form.

9.5.2 Inspection items and intervals

Adequate inspecting intervals are allocated to autonomous maintenance with reference to shift, day, week, and month. Because operators spend a considerably long time in the preparation and clearance of rou-

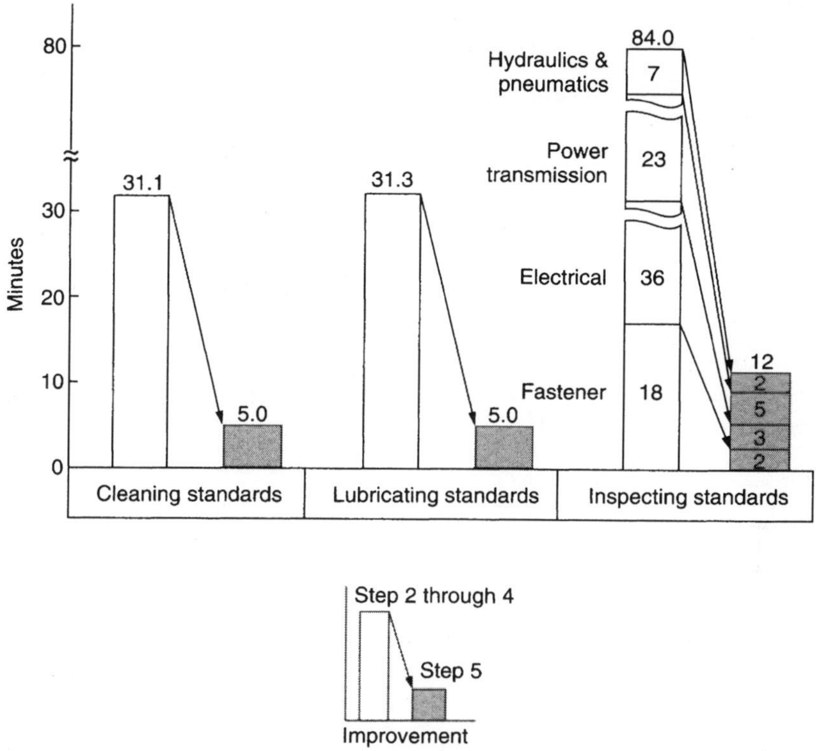

Figure 9.4 A reduction in the time spent for operators' routine maintenance.

tine tasks as well as the cleaning and lubricating designated in relevant standards, it is difficult to spare much additional time for inspection. Therefore, the inspection scheduled per shift or day is restricted to minimal items which prevent severe troubles in terms of safety and quality. Other inspections are be performed on a weekly or monthly basis.

With the above considerations in mind, operators must master all of these inspections as "conditioned reflexes," rather than depending on check sheets widely used in typical factories. It is, on the other hand, quite unreasonable to expect operators to inspect daily too many items which cannot be conducted without the aid of check sheets.

It is, therefore, better to provide adequate time for inspecting fewer items by extending intervals rather than more items at shorter intervals. For instance, twenty minutes spent at the end of every month is more effective than two extra minutes a day. With this more feasible approach to the expansion of time, operators definitely conduct more thorough and reliable inspections.

TABLE 9.2 Inspection Standards Prepared by Full-Time Maintenance (Left half of a sheet of paper)

Prep'd by				Periodical			**Inspection standards**									
App'd by				Routine												
Scopes									Departments				Intervals			
Safety	Quality	Equipment	Subassembly	No.	Location	Inspection items	Uptime	Downtime	Maint.	Prod.	P.E.	Subcon.	Year	Month	Week	Day
		○	Rubber cutter		Cutting blade	Wear & crack	○			○				1		
		○			Cylinder	Actuation	○		○					3		
		○			Holder	Deformation & breakage	○		○					3		
○					Cover	Deformation & breakage	○		○					3		
	○		Chuck		Roller	Deformation & rotation	○		○					3		
		○			Cylinder	Actuation	○		○					3		
	○				Bead support	Deformation	○		○					3		
		○	Back feeder guide		Back stop	Deformation & abrasion	○		○					3		
		○			Cylinder (Back stop)	Actuation	○		○					3		
	○				Height	Gap with cutter		○	○					3		

TABLE 9.2 *(Continued)*

		❍			Upper & lower cylinder	Actuation	❍		❍					3		
		❍			Center stop roller	Deformation & abrasion	❍		❍					3		
		❍			Drop prevention hook	Deformation & abrasion	❍		❍					3		
		❍			Guide roller	Deformation & rotation	❍		❍					3		
❍			Front cover		Cover	Damage	❍		❍					3		
	❍		Flat stop		Roller	Deformation & breakage	❍		❍					3		
	❍				Cylinder	Actuation	❍		❍					3		
		❍	Common		Pneumatic devices & pipes	Air leakage	❍		❍					3		
		❍			Sensors	Installation conditions	❍		❍					3		

Revision summary			Specification
Page	Rev. No	Date revised	Spec. No.

TABLE 9.2 Inspection Standards Prepared by Full-Time Maintenance (Right half of a sheet of paper) *(Continued)*

❍ Mechanical standards Instrumental standards	Copy to	Applicable codes

Inspection methods	Corrective actions	Notes
Check cutting shape of rubber materials.	Sharpen or replace	
Visually inspect smooth actuation with 6.0 kg/cm^2 air.	Adjust or repair	
Visually inspect defective parts.	Repair	
Visually inspect defective parts.	Replace	
Visually inspect defective parts and confirm smooth rotation.	Replace	
Confirm smooth motion with 3.5 ~ 4.0 kg/cm^2 air.	Adjust or replace	
Visually inspect defective parts.	Repair or replace	
Visually inspect defective parts.	Replace	
Confirm smooth motion.	Repair	
Check 0.1 ~ 0.3 mm gap with thickness gauge.	Adjust gap	

TABLE 9.2 *(Continued)*

Confirm smooth motion with 3.5 ~ 4.0 kg/cm^2 air.	Adjust or repair		
Check edge misalignments by actual materials.	Adjust		
Confirm nonchafing in rubber materials.	Replace		
Visually inspect defective parts and confirm smooth motion.	Replace or adjust		
Visually inspect defective parts.	Repair or replace		
Visually inspect defective parts.	Repair or replace		
Confirm smooth motion.	Repair or replace		
Check leakage with soapy water.	Repair or replace		
Check loose installation.	Tighten		
Equipment	Rubber cutting machine	Mechanical	
Applicable		Electrical	

Furthermore, as operators become increasingly skilled in discovering minute defects in equipment during routine tasks, fewer formal inspections are required. It is this approach to inspection that is originally anticipated for operators from the beginning of the TPM program.

9.5.3 Inspection time

When inspecting items and intervals are set for routine inspection, the time allocated is roughly estimated. On the other hand, the actual length of time required for inspection varies according to the equipment and operating conditions involved. In general, inspection times are decided based on the following:

- Do the operator's routine tasks consist of process condition monitoring, machine operations, or repetitive manual work?
- How many machines does an operator manage?
- What work motions and routes does an operator follow?
- To what extent is equipment automated?
- Does process capacity have adequate allowances?
- Does an operator inspect equipment during up- or downtime?

In reality, almost all the time given for inspection is determined by how much time can be extracted from the operator's actual work hours. First of all, taking the above conditions into consideration, tentative time targets are set. Then trial inspections are carried out for each item at varying intervals.

Noting the differences between targeted and expended times, operators improve inspecting methods and equipment, and review the inspection intervals and allocations of tasks for the production and maintenance departments. Based on this trial and error approach, optimal combinations are discovered.

Although it is highly time consuming in the early phase, inspection time can be reduced to some extent by repetitive experience. It is important to review carefully the combined routine maintenance activities of cleaning, lubricating, and inspecting, in order to inspect equipment efficiently while it is being cleaned and lubricated.

9.6 Preparing Autonomous Maintenance Standards

9.6.1 Make a suitable combination of cleaning, lubrication, and inspection

In order to prepare efficient standards, it is necessary to lay out a suitable combination of the operator's work sequence and route.

Find an optimum combination. No operator should need to walk around the same process area three times to achieve cleaning, lubricating, and inspecting. Once work items and intervals are fixed, operators can determine what must be done on each shift or day and at which particular areas of equipment. Operators consider matters such as, "Is the assignment of single or multiple operators better?," "Are both hands free?," "Can operators carry necessary tools in their hands?," "Is the combination of these routine tasks adequate?" As a result, various combinations are tried:

- Cleaning, lubricating, or inspecting as separate items
- Cleaning and lubricating
- Cleaning and inspecting
- Inspecting and lubricating
- Cleaning and lubricating and inspecting

Decide upon operators' work sequences and routes. During the activities through Step 4, operators determined the sequence of their tasks and the route to reach given time targets when they prepared cleaning, lubricating, and inspecting standards. Certain sequences and routes, however, are not always the best to follow when executing routine maintenance in the combinations just described. Operators, therefore, discover the optimum route by rearranging their work patterns and movements.

9.6.2 How to establish autonomous maintenance standards

Two examples of maintenance work standards are introduced in Tables 7.5 and 9.3. Basically, the details listed here are prescribed in autonomous maintenance standards according to the operator's work sequences and route.

Cleaning standards

- Location: Parts, subassembly, equipment, and process
- Criteria: Free from dirt, dust, waste oil, or metal chips
- Method: Sweep up, wipe off, absorb, or drain
- Tools: Rag, broom, brush, scraper, vacuum cleaner, or specific utensil

Lubricating standards

- Location: Parts, subassembly, and equipment
- Criteria: Within the levels as indicated at lubricating points, or resulting sufficient oil film
- Lubricants: Type, viscosity, and code number
- Tool: Hand lubricator, pump, or pressure gun

TABLE 9.3 Autonomous Maintenance Standards

Autonomous maintenance (cleaning, lubricating & inspecting) standards				Prepared by ______ Date issued ______					
Process: No. 1 Rubber product plant, No. 3 Rubber molding		Equipment: TR-62		Approved by ______ Date expired ______					
	No.	Cleaning areas	Criteria	Methods	Min.	Interval			Person responsible
						Day	Week	Month	
Cleaning	1	Main body of injection molding machine	No abnormal contamination by grease, dirt and dust	Wipe with damp rag	10		○		Operator
	2	Main body of press	ditto	ditto	10			○	ditto
	3	Hydraulic unit	ditto	ditto	5		○		ditto
	4	Temperature controller	ditto	ditto	2		○		ditto
	5	Mold disassembly jig	Rubber flash and chips within limits	Sweep with broom	5	○			ditto
	6	Coater	No unfavorable effects on quality and sliding parts by ink residue	Remove ink residue with scraper and brush or replace jig	5		○		ditto
	7	Underground pit	No oil leakage or cooling water overflow	Eject with vacuum pump	3			○	ditto

TABLE 9.3 *(Continued)*

	No.	Lubrication points	Criteria	Lubricant types	Tools	Min.	Interval			Person responsible
							Day	Week	Month	
Lubricating	8	Hydraulic oil reservoir	Oil level within ranges marked	Yellow-56	Hand pump	1			○	Operator
	9	Sliding base in injection molding machine	Sufficient oil film	Blue-68	Oiler	1			○	ditto
	10	Air lubricator	Oil level within ranges marked	Yellow-32	ditto	1		○		ditto
	11	Heating plate cooling water tank	ditto	Green-68	ditto	1			○	ditto
	12	Centralized greasing unit	ditto	White-B-1	Pressure gun	1		○		ditto
	13	Screw gear in hydraulic clutch	ditto	Green-32	Oiler	1			○	ditto
	14	Rotating parts in press	Sufficient grease	White-C-1	Pressure gun	3			○	ditto
	15	Reservoir in temperature controller	Oil level within ranges marked	Green-68	Oil container	1		○		ditto

TABLE 9.3 Autonomous Maintenance Standards (*Continued*)

	No.	Category	Location	Criteria	Methods	Corrective actions	Min.	Interval			Person responsible
								Day	Week	Month	
Inspecting	16	Hydraulics	Oil pressure in temperature controller	1.0~2.0 kg/cm^2	Visual	Adjust	1	❍			Operator
	17	Hydraulics	Oil temperature in hydraulic system	35°C~55°C	ditto	Shut down & call maintenance dept.	1	❍			ditto
	18	Hydraulics	Hydraulic pump indicator	Within "filter clean" position	ditto	ditto	1	❍			ditto
	19	Fastener	Die holder bolts	No looseness	Tap lightly	Tighten	6	❍			ditto
	20	Power transmission	Bearing wear in mold disassembly jig	No wear	Visual	Replace	1	❍			ditto
	21	Power transmission	Fan belt	ditto	ditto	ditto	1		❍		ditto
	22	Electrical	Safety door limit switch	No looseness in setting bolts	ditto	Tighten	1	❍			ditto
		Electrical	Lamps at console and machines	No burnout or damage	ditto	Replace	1	❍			ditto
	23	Hydraulics/pneumatics	Pipes and fittings	No leakage or damage	ditto	Call maintenance dept.	5	❍			ditto

Inspecting standards

- Category: Fastener, electrical, power transmission, hydraulics, or pneumatics.
- Location: Parts, subassembly, equipment, and process.
- Criteria: Sample, reading of instruments, color, noise, vibration, overheating, odor.
- Method: Visually observe, touch, listen, smell, read instruments.
- Corrective action: Adjust, replace parts, tighten, drain, or call maintenance personnel.

In addition to the above, common items, such as time frame, intervals, and names of personnel in charge, are recorded in writing. All operators are required to master necessary skills; otherwise, the column for listing responsible personnel cannot be filled in with "operator." Operators with sufficient skill are placed in charge of particular inspections not addressed in the preceding steps or inspections requiring the use of specific instrumental apparatus.

Utmost efforts must, however, be made in order to bring all operators to this higher level of proficiency as early as possible. It also is important to specify clearly the effective limits of standards to ensure the review and obligatory revision of the standards concerned.

Illustrate standards as simply as possible. Typical examples of cleaning/lubricating and autonomous maintenance standards are presented in Chaps. 7 and 9. Operators, however, must illustrate these standards as simply as possible by way of suitable aids, such as a routine maintenance map similar to the drawing shown in Fig. 7.4 and instructive labels displayed on each spot of equipment.

9.7 Case Study

9.7.1 A routine inspection control board

In a factory, routine inspection control boards are utilized to prevent the omission or delay of necessary inspections, as illustrated in Fig. 9.5. Anyone can ascertain immediately the progress of inspection by referring to these boards, which represent an application of visual control.

9.7.2 Breakdown prevention in a performance tester

In an assembly line, an operator detected an unusual noise originating from the main shaft of a performance tester and informed the mainte-

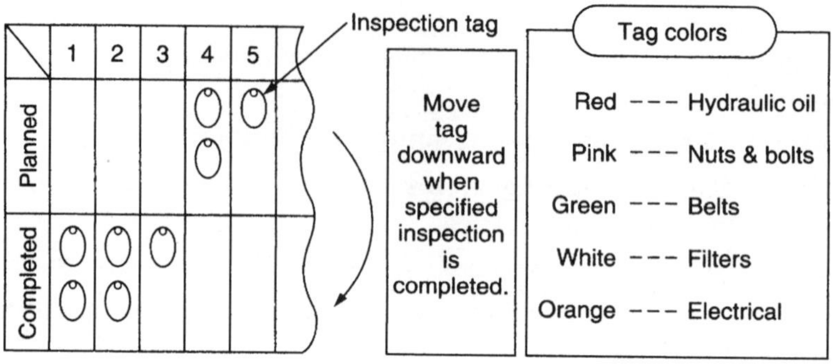

Figure 9.5 A control board for routine maintenance (visual control).

nance department about this condition with a "yellow card," as illustrated in Fig. 9.1.

As the consequence of a disassembling inspection, maintenance personnel found that the noise was caused by abrasion in a keyway of the shaft. Fabrication of the required replacement parts was immediately arranged for and the main shaft was replaced during the following weekend maintenance session, as illustrated in Fig. 9.6.

Figures 9.7 and 9.8 show the number of abnormalities detected and reported to the maintenance department by the type of indications, and the increasing number of warning cards issued, which resulted in the effective prevention of further damages.

9.7.3 Breakdown prevention in a pin milling machine

In another plant, a "yellow card" system also resulted in remarkable effects. For example, an operator detected abnormal noise in a pin milling machine installed in a C-type crankshaft production line. According to a machine diagnosis conducted by maintenance personnel, a comparison between problematic No. 4 and normal No. 2 stations, as delineated in Table 9.4, ascertained that the noise was caused by abrasion in a bearing unit. A serious breakdown, as a result, was prevented before it occurred.

When the maintenance department receives a warning card from the production department, an advantageous result is the application of subsequent inspections to similar machines. Figure 9.9 shows the results of inspecting 244 other similar machines in response to 61 warning cards reported by operators during a half-year period.

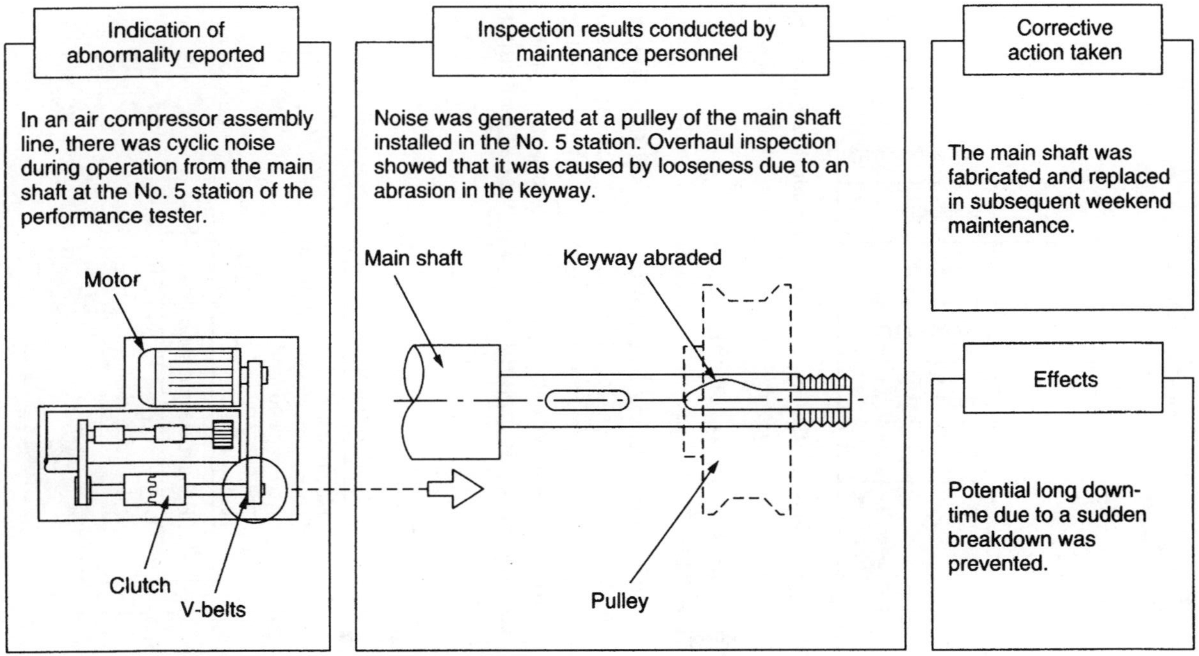

Figure 9.6 Preventing serious breakdowns in performance testers.

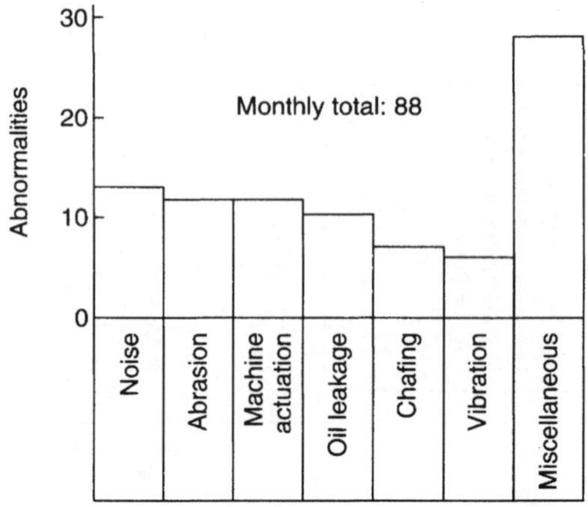

Figure 9.7 Classifying abnormalities by the types of indications discovered over a month.

9.8 The Keypoints of an Autonomous Maintenance Audit

To finish the activities relating to equipment, auditors check that all issues are completed adequately. If some issues are not resolved yet, operators must plan concrete countermeasures. In reviewing Steps 1 through 5, auditors check whether operators are pursuing the causes of breakdowns and minor stoppages, and that visual controls are thoroughly implemented.

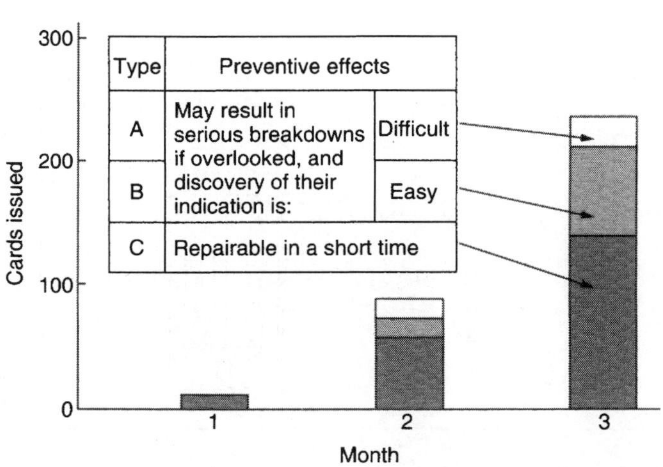

Figure 9.8 The growth of effective yellow cards in the prevention of sudden breakdowns.

TABLE 9.4 Preventing Breakdowns at Pin Milling Machines

Inspection item	No. 4	No. 2
Vibration (A-DBV)	4.3	1.7
Maximum current (A)	200	150
Oil temperature (°C)	47	45
Oil contamination	High	None
Noise (dB)	93.4	88.2
Judgment	Abnormal	Normal

In addition, each operator must show that he or she can clean, lubricate, and inspect equipment within targeted times in accordance with the autonomous maintenance standards. Auditors confirm that the standards are feasible to follow, and that they are actually followed by the operators. Auditors also need to determine whether operators definitely adopt methods for effective improvement, as achieved in other processes applicable to their own similar equipment.

Even when Step 5 is successfully completed, it is not the end of activities focused on equipment. To the contrary, higher technical levels already achieved by operators and maintenance personnel allow for better powers of observation directed toward equipment. As a result, an increased awareness of new issues develops and these matters are continuously dealt with as subjects for short remedial programs.

The well-established implementation of autonomous supervision founded on the CAPD cycle is a prerequisite when addressing matters of quality, which are much more difficult than breakdowns. In the autonomous maintenance audit, it is most important not only to confirm equipment conditions, the finished writing of autonomous maintenance standards, and the feasibility of time targets, but also to

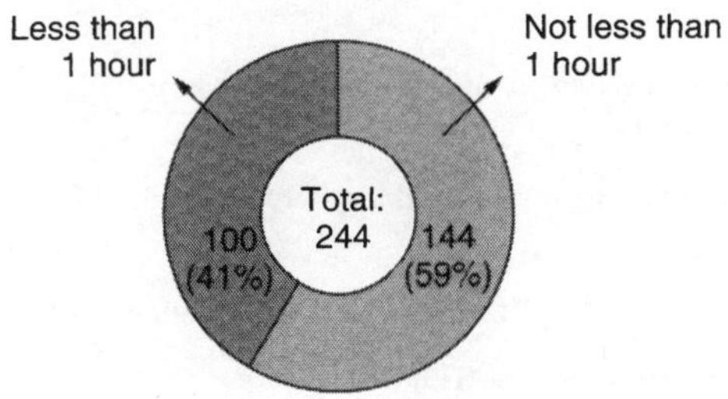

Figure 9.9 The results of an inspection for other similar machines.

TABLE 9.5

Step 5	Autonomous Maintenance Audit Sheets	Sheet 1 of 3			
No.	Audit points	Results			
1.	**Steps 1 through 4 review**				
1.1	Are equipment defects detected from Step 1 through 4 definitely corrected?				
1.2	Are remedial actions to sources of contamination completed?				
1.3	Are difficult areas for conducting routine maintenance adequately modified?				
1.4	No residual issues in above? If so, do reasonable explanations exist? Are plans and schedules to solve these issues clear?				
1.5	No unanswered questions?				
2.	**Group activity** (General)				
2.1	Are aims of Step 5 understood adequately?				
2.2	Is activity plan made in advance? Well executed?				
2.3	Are managers' models well understood?				
2.4	Is activity board adequately utilized?				
2.5	Are safety matters carefully respected?				
2.6	Are TPM activity hours and frequency adequate?				
2.7	Is more efficient way of TPM activity pursued?				
2.8	Are used spare parts and consumables recorded?				
2.9	Is meeting after on-site activity definitely held? Reports submitted?				
2.10	Is activity participated in by all members? No indication of dropout?				
2.11	Are all members cooperating equally? Not led by particular member?				
2.12	Are noteworthy ideas introduced actively to other PM groups?				
2.13	Is cooperation with full-time maintenance satisfactory?				
3.	**Routine maintenance activity**				
3.1	Are cleaning, lubricating and inspecting times sufficiently reduced? Are relevant standards faithfully revised whenever work methods or equipment are modified?				

TABLE 9.5 *(Continued)*

Step 5	Autonomous Maintenance Audit Sheets	Sheet	2 of	3	
3.2	Are breakdowns caused by internal deterioration prevented with the five senses? Is warning card system effectively operated?				
3.3	Are countermeasures against misoperation taken in cooperation with full-time maintenance?				
3.4	Are remarkable cases of breakdowns and misoperation prevention actively taught to all PM groups? Are related standards always reviewed?				
3.5	Are standards prepared in the past steps sufficiently assessed in terms of safety, quality and breakdowns? Are discovered problems remedied?				
3.6	Are tentative inspecting standards set in Step 4 thoroughly reviewed by making a detailed comparison with those prepared by full-time maintenance? Is allocation of inspecting tasks between autonomous and full-time maintenance adequate?				
3.7	Are autonomous maintenance standards, by systematic-ally combining various standards, set in the past steps?				
3.8	Are maintenance schedules prepared to definitely conduct operator's routine maintenance? Faithfully executed?				
3.9	Is routine maintenance work to be done during up/down-time of equipment clearly distinguished? Well understood?				
3.10	Can everyone clean, lubricate and inspect equipment in accordance with autonomous maintenance standards? Time targets achieved? If not, are adequate plans and schedules prepared?				
4.	**Equipment** (Main body and surroundings)				
4.1	Are toods and jigs stored at designated locations? No shortage or damage?				
4.2	No unnecessary pieces of equipment?				
4.3	No unnecessary materials in process? No dropped parts on floor?				
4.4	Can quality/defective products or scraps be clearly distinguished?				
4.5	Do status display and warning lamps function properly?				
4.6	Do safety devices function properly?				

TABLE 9.5 *(Continued)*

Step 5	Autonomous Maintenance Audit Sheets	Sheet 3 of 3			
5.	**Visual control**				
5.1	Are visual controls thoroughly applied to facilitate inspecting tasks?				
5.2	Are new visual controls devised? Noteworthy ideas actively revealed to other PM groups?				
6.	**Short remedial program**				
6.1	Is subject selected from the six big losses?				
6.2	Are matters of quality being dealt with?				
6.3	Are problems clearly identified? Targets pinpointed?				
6.4	No easy countermeasures taken?				
6.5	Are cost and effects of program reviewed? Actual figures recorded?				
6.6	Are preventive measures against recurrence of problems provided?				
7.	**Residual issues**				
7.1	No residual equipment related issues? If so, do reasonable explanations exist? Are plans and schedules to solve these issues clear? Is adequate assistance of full-time maintenance arranged?				

determine with certainty whether autonomous supervision is fully implemented in PM groups.

An evaluation of these matters, of course, is not achieved in the limited time set for the on-site audit and the subsequent meeting. In addition to careful daily supervision and observation, managers must arrive at and finally evaluate the conclusions of the step audit. If some PM groups have any weak points, in spite of a satisfactory audit, managers must effectively instruct operators on how to remedy them in Step 6. Table 9.5 shows a sample of audit sheets.

Chapter 10

Step 6: Process Quality Assurance

All activities completed in Steps 1 through 5 focus on equipment and Zero Breakdowns. In Step 6, the autonomous maintenance activities are directed toward the aim of Zero Defects. Product quality must be assured not by unpredictable human behavior, but by a process to reduce quality defects to Zero. In this chapter, the term "process" generally refers to any functional unit consisting of a single or multiple pieces of equipment built into a production line.

10.1 Aims from the Equipment Perspective

A highly reliable process that manufactures only quality products at all times is realized by means of a clear identification of process quality and the related upkeep of quality conditions. When this is not possible, a process that does not result in the outflow of defective products to downstream processes involving the outside of the factory may be achieved by improving equipment as well as work methods, with various aids such as visual controls and mistake proofs.

Step 6 aims at the realization of a truly orderly shopfloor, from the viewpoint of process quality assurance, where not only equipment, but also any kind of materials which exist throughout the factory, is determined by anyone to be normal or abnormal at a glance.

10.2 Aims from the Human Perspective

By comprehending the working mechanism of machinery, along with the basic concepts and methods of quality assurance, operators develop two major activities: first of all, preventing outflow of defective products to downstream processes, and, second, preventing the manufacturing of defective products themselves. Step 6 aims to foster the development of operators knowledgeable in matters of quality, equipment, and the procedure for fully implementing autonomous supervision, so as to eliminate the need for detailed instructions by managers.

10.3 The Challenge of Zero Defects

10.3.1 Recent trends

In keeping with the rapid spread and deep penetration of the TPM concept recently developed in various types of industry, targeted goals of TPM activities are becoming more ambitious year by year. No one is surprised to hear that the occurrence of breakdowns is reduced to one-tenth or one-twentieth of bench marks measured prior to the implementation of TPM. It may be a common assumption in Japan that breakdowns can be reduced to such levels when serious endeavors to do so are made according to the TPM concept.

Meanwhile, when minor stoppages and breakdowns are drastically reduced by thoroughly eliminating deteriorated and defective parts in equipment, the occurrence of quality defects, as a natural consequence, may also be considerably reduced. An assumption that complete satisfaction will be achieved in this way is, however, not an appropriate approach to the goal of Zero Defects. The most successful companies are more interested in matters of quality rather than of breakdowns per se, because the attainment of Zero Defects is much more difficult than that of Zero Breakdowns.

In the case of breakdowns, it is possible to detect their sporadic occurrence by indications such as abnormal vibration, noise, overheating, and odor. Besides, the causes of breakdowns have a certain broad tolerance. On the other hand, the causes of quality defects have much narrower tolerances and a larger variation in types of occurrence. In a process which manufactures solid products, the following factors affect quality:

- Accuracy, material and hardness provided for parts of equipment, dies, tools, and measuring instruments
- Physical property of raw materials and additives
- Positioning and machining methods
- Operating conditions of equipment and work methods

In addition, quality is frequently influenced by changes in circumstances like atmospheric temperature, humidity, and airflow. Thermal expansion or a slight change in frictional force sometimes have no influence on breakdowns, but do affect quality.

Accordingly, after attaining Zero Breakdowns in Steps 1 through 5, frontline personnel must then strive for qualitative matters in Step 6 to attain Zero Defects within several years' time. This approach becomes even more important in view of the progress of automation among the assembly industries in recent years. Prior to a detailed discussion in these regards, the TPM approach to quality is described in the following sections.

10.3.2 Process quality assurance

Assuming that raw materials or component parts coming from upstream processes meet quality requirements, quality in a given process is determined by operating conditions of machinery, human beings, and their work methods.

In a traditional machining process of metal products, for example, operators used to cut, grind, press, and weld materials by manipulating machine tools installed along a conveyor. What was requested of operators in those days were the skills required to manipulate machine tools by hand. At the same time, there were many, various types of simple, repetitive work, such as workpiece setting and removal, material handling, cleaning, and so on. Operators, however, were able to manufacture quality products by their craftsmanship in spite of inferior machine tools.

At that time, the human role used to be the principal and direct determiner of quality. Quality was created by human beings, work methods, and machinery, in that order. On the other hand, automation of the assembly industry has rapidly accelerated since the late 1970s. Automation was applied especially to those processes in which it reduced production costs, improved quality, and lent itself to easy implementation.

Machinery, however, is not as flexible as human beings. When the workers used to set workpieces on a machine, it was considered satisfactorily done if no workpiece dropped from a conveyor as it moved on from worker to worker. Once the setting of workpieces are automated and workers are replaced by robots, these robots cannot pick up workpieces unless they are located always in the same position, direction, and posture within a certain narrow tolerance.

In recent decades, these factors which are intrinsic to equipment and directly affect quality, have rapidly increased in number and types every year in keeping with the pace of automation of manual work. Today, what creates quality is primarily the operating conditions of processes. Now, quality is influenced by machinery, work methods, and human beings, in this new order.

In these automated factories, the operators' role is shifting to the operating, monitoring, and managing of equipment or processes, and only indirectly relates to quality. Consequently, requirements for manipulating machine tools by hand are fading away. In other words, automated processes require operators to be knowledgeable about equipment, its operating conditions, and qualitative matters, rather than to be experts in the manipulation of machine tools.

In TPM, process quality assurance refers to the proper maintenance of a piece of equipment or process, as shown in Fig. 10.1, with the objective of Zero Defects by concentrating on the relationship between equipment conditions and product quality.

10.3.3 The nature of quality conditions

There are two types of subassemblies which make up a piece of equipment. One has a primary function such as machining, coating, assembling, inspecting, packing, and conveying materials. The other provides auxiliary functions in support of the primary functions, such as driving, power transmission, lubricating, cooling, and control. Most pieces of equipment consist of these two types of subassemblies, designed to be as independent and separate as possible from each other to secure higher maintainability, operability, and reliability.

Of course, quality defects will still occur, unless these two types of functions are maintained properly. In particular, those subassemblies

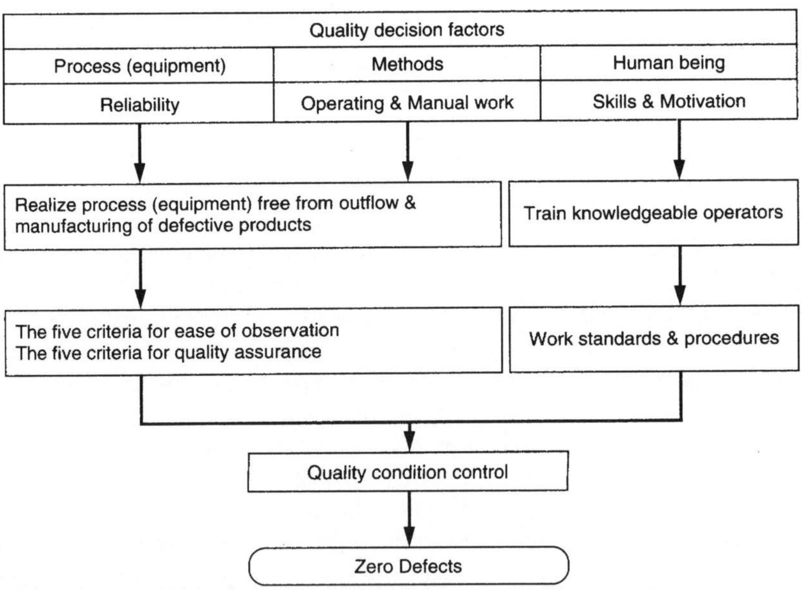

Figure 10.1 An approach to Zero Defects.

which directly contact workpieces or supply additives, like paint and adhesive, have critical and essential influences on quality. Generally, multiple factors that determine quality are built into a piece of equipment or subassembly; for example:

- Shaft, bearing, and hydraulic systems which rotate or reciprocate workpieces
- Positioning pins, reference planes, chucks, and clamps which fix or position workpieces
- Cutting tools and dies which cut or punch workpieces
- In plastic extrusion molding process, operating temperature, cleanliness of surfaces of molds, pressure screws, and interior of heating chambers
- In car painting or metal decorating process, air flow and temperature distribution in oven

Each single factor mentioned above involves a certain number of conditions that create specified quality. These conditions are referred to as "quality conditions." If quality conditions are maintained properly at all times, no defective or off-specification products will be manufactured. In other words, process quality assurance is equal with the proper maintenance of quality determination factors of equipment or quality conditions.

There are the following two approaches to achieving quality conditions:

1. An *abstract approach* carried out by the product design and plant engineering departments during the engineering stage. By making a detailed analysis of the quality of a future product to be manufactured by future equipment, quality specifications and the resulting quality conditions are determined.
2. A *concrete approach* carried out by the production, maintenance, and plant engineering departments throughout the commercial production stage. By making careful observations on the operating conditions of equipment and resulting quality, the quality created by existing equipment is analyzed concurrently with the clear identification of quality conditions.

In both approaches, the plant engineering department must possess a high level of capability for analyzing and designing quality conditions, and also must assist the related departments in various technical aspects.

To make discussion easier, this book is focused on the one-piece flow production of solid products. This basic approach, however, is quite applicable to continuous processing, such as in the iron manufacturing, aluminum refining, paper making, and plastic molding industries, or any other process industries which deal with fluid, powder, or other bulk materials. Most quality conditions in the assembly industries relate to mechanical conditions. On the contrary, in the process industries, quality conditions consist of various physical properties, for example, temperature, pressure, flow volume, viscosity, and specific gravity, and, therefore, different process control systems and measuring instruments are required.

10.3.4 The relationship among quality specifications, quality causes, and quality results

The relationship between equipment and quality in accordance with three aspects of quality—quality specifications, quality causes, and quality results—is defined here and diagrammed in Fig. 10.2.

Quality specifications: In the engineering stage of a new product, the accurate details of quality are determined when the particular design of a new product is fixed; for example, material, shape, dimension, accuracy, surface finish, color, and coating thickness. The details of quality to be created by future equipment are referred to as "quality specifications."

Quality causes: Equipment (quality conditions) to manufacture a new product which meets the prescribed quality specifications is designed within a given budgetary limitation. At the time when the equipment is handed over to the production and maintenance departments after construction and commissioning, factors incorporated in equipment to create quality of a new product are fixed. These factors are referred to as "quality causes."

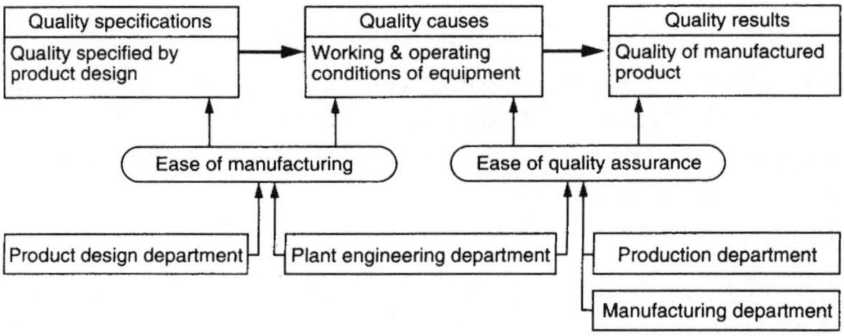

Figure 10.2 The relationship among quality specifications, quality causes, and quality results.

Quality results: During the commercial production stage, operating and maintaining conditions of equipment (quality conditions) determine quality. "Quality results" refer to the quality of the product manufactured which is created by existing equipment.

Most of the equipment installed in a plant is procured from vendors. Naturally, its quality causes are obscure in order to satisfy various users' wide spectrum of needs. When such equipment is purchased, plant engineers must be able to incorporate their own requirements for quality into this equipment as early as possible at any of the engineering, procurement, fabrication, or construction stages. A company's ability in this regard reveals its technical level as an equipment user.

On the other hand, the test run and commissioning period is the time to check whether the quality causes designed by plant engineers actually meet the product design requirements. In other words, quality results are verified in terms of conformance to quality specifications. In general, because quality causes may involve many design mistakes and inadequate anticipations, they do not easily coincide with quality specifications. Sometimes, design mistakes are discovered in quality specifications and incorporated in product design, even though a prototype was made in the engineering stage.

These troubles, as a consequence, prolong the commissioning period and postpone the initiation of commercial production, in other words, handover of the plant to the production and maintenance departments. In many factories, a large number of quality defects are occurring every day due to poor technical capacity and a lack of serious efforts directed toward matters of quality. In order to attain Zero Defects, the greatest effort is required of the production department as well as all other quality related departments, such as maintenance, plant engineering, product design, quality assurance, and so on.

10.3.5 The five quality assurance criteria

Assuming that proper quality causes incorporated into equipment are turned over by the plant engineering department, management of quality causes, in other words, subsequent process quality assurance, is the task of the production as well as maintenance department by way of proper operation and maintenance of equipment.

At this point, a question may be posed: "Are there some effective criteria by which quality conditions may be assured from the engineering stage on through routine production?" Because quality conditions were not clearly set in the engineering stage and remain obscure or unstable in the commercial production stage, quality defects occur frequently and chronically. To understand the relationship between quality conditions and the recurrence of quality defects, the following examples are presented:

Quality conditions are unclear

- Operators as well as maintenance personnel do not know very precisely what the quality conditions are nor how to manage these conditions and at which areas of equipment.
- Quality conditions have too wide, too narrow, or ambiguous ranges.
- Quality conditions are adjusted by relying on operators' experiences or feelings.
- When equipment is started up, adjustments and test runs must be repeated.

Quality conditions are difficult to set

- Procedure for setting quality conditions is difficult or complicated.
- Setup and adjustment requires too much labor and time.

Quality conditions tend to vary

- Quality defects readily recur in spite of frequent adjustment of operating conditions.
- Adjustment and fixing of operating conditions must be repeated many times.

Quality conditions are difficult to detect

- Changes in quality conditions are not found until a large number of defective products are already produced.
- It is difficult to identify which condition changed.

Quality conditions are difficult to restore

- Disassemblage of equipment is needed to set quality conditions and, therefore, involves excessive labor and time.
- Welding and file finishing are needed to restore quality conditions of worn parts and require that a work order be issued to the maintenance department.

According to the above considerations of the relationship between these occurrences of defective products and concurrent quality conditions, new criteria to assure quality in connection with a specific piece of equipment or process should be introduced. These criteria are referred to as "the five criteria for quality assurance" and are described here:

1. *A quality condition is quantitative or clear.*
2. *A quality condition is easy to set.*

3. *A quality condition resists variation.*
4. *A change in a quality condition is easy to detect.*
5. *A change in a quality condition is easy to restore.*

An example of an assembly process associated with the installation of a slide door in a commercial car is described here and illustrated in Fig. 10.3.[1] The accurate setting of a gap between the body and door is one of the important items for quality assurance in terms of airtightness and waterproofing. Numbers (1) to (6) of Fig. 10.3 indicate the measuring positions for accuracy of fit. Figure. 10.4 follows the trend of measured accuracy. According to the charts, the accuracy of door assembly presented no problem in the early stage of commercial production. The accuracy did, however, begin to deteriorate gradually approximately eight months after start-up of the car assembly line.

A detailed survey revealed that these quality defects recurring in a body subassembly process were caused by worn parts of a positioning jig installed therein. The accuracy of these parts was the most critical quality condition in this process; for example, the given operating standards specified that the reference stops should be inspected only once a year by using a height gauge and the abrasion allowance must be less than 0.3 mm.

It is instructive to scrutinize the quality conditions incorporated into these stops when they were designed in light of the above-mentioned five criteria for quality assurance as listed here:

- Abrasive material is selected. — Criterion 3: Quality condition resists variation.
- Abrasion is not easily monitored due to a troublesome and inaccurate measuring method. — Criterion 4: Change in quality condition is easy to detect.

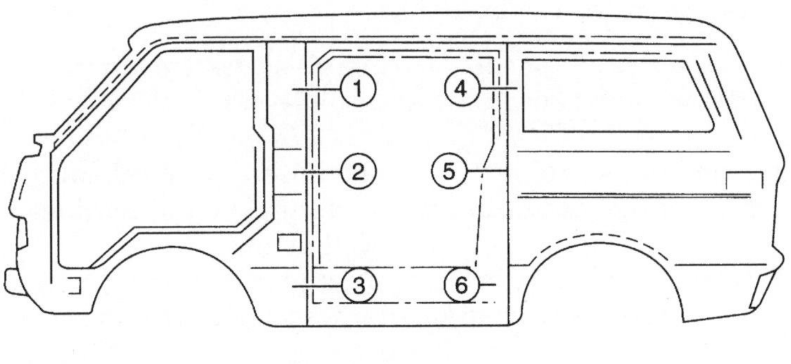

Figure 10.3 The accuracy of measuring positions.

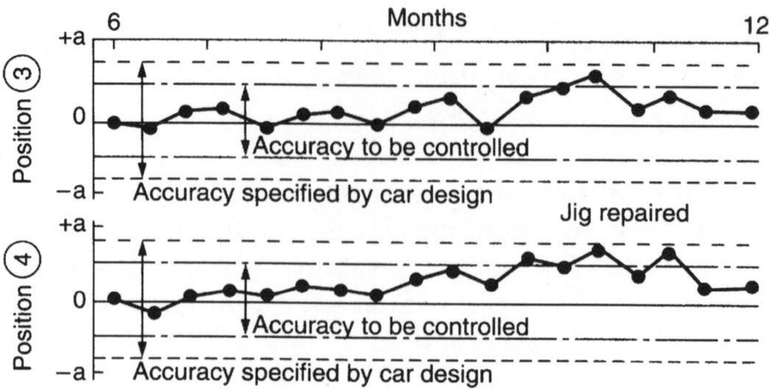

Figure 10.4 The trend of accuracy measured at positions ③ and ④ of Fig. 10.3.

- Worn stops are restored by overlay welding and file finishing. — Criterion 5: Change in quality condition is easy to restore.

It then became clear that three criteria out of the five were not satisfied in the existing jig, resulting in the deterioration of the quality of door fitness. As a result of the above assessment, the said stops were modified to meet all of the five criteria for quality assurance as described below:

Criterion 3: Material and surface treatment of the stops is changed to higher hardness, as shown in Fig. 10.5.

Criterion 4: Reference stops and positioning pins are punch marked with a depth equivalent to the specified tolerances for door fit defined in the car design specifications as illustrated in Fig. 10.5. When the punch mark disappears, as is eventually likely, anyone can identify, at a glance, that the part wear has reached its limit of abrasion allowance, as illustrated in Fig. 10.6.

Criterion 5: Reference stops are modified to be removable by being fixed onto the jig with bolts, so as to facilitate quick replacement and eliminate necessity for any specific skill, such as welding and file finishing, as illustrated in Fig. 10.5.

Thanks to a series of remedial actions, all of the five criteria for quality assurance were satisfied and, eventually, a highly reliable jig was realized.

Quality conditions are, however, not limited to mechanical conditions, such as parts shape, dimension, accuracy, and hardness. They include other physical properties as well as electrical characteristics. Although setting, adjusting, measuring, and maintaining methods differ with characteristics to be assured, these same criteria are applica-

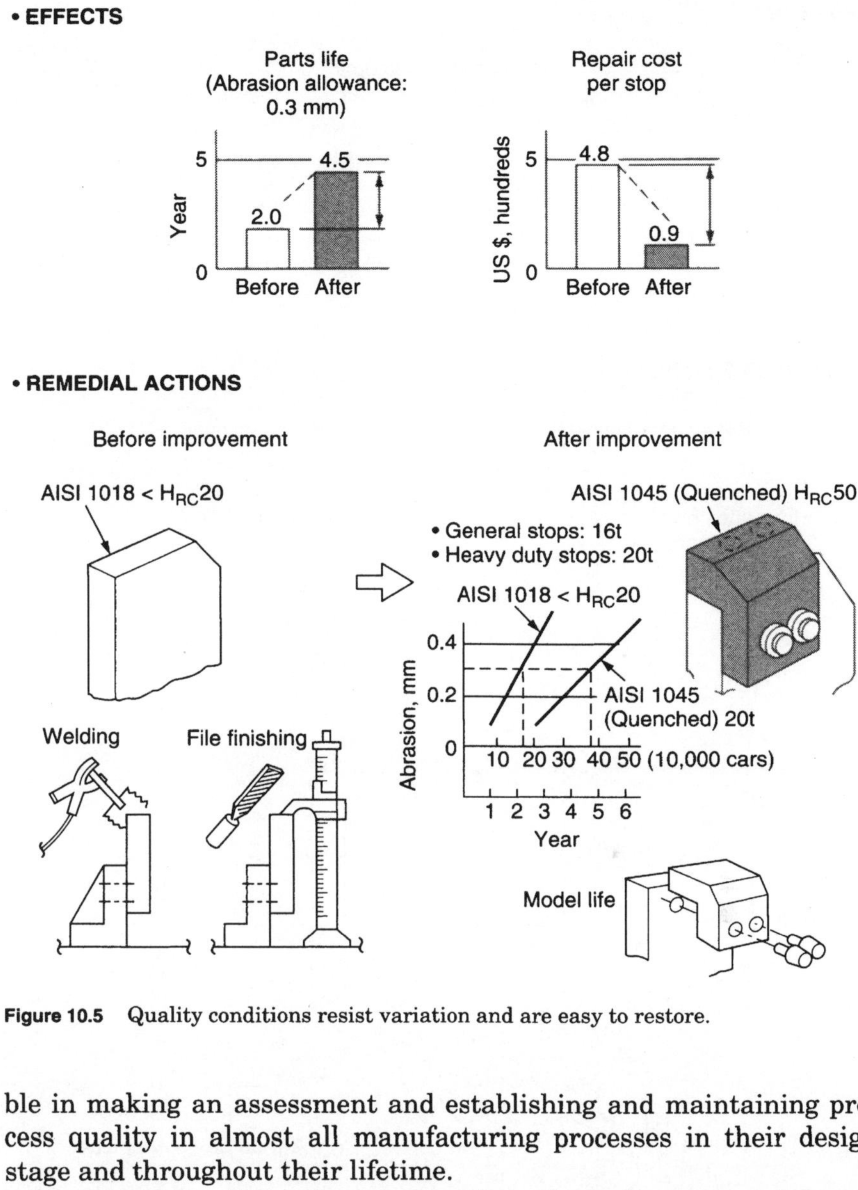

Figure 10.5 Quality conditions resist variation and are easy to restore.

ble in making an assessment and establishing and maintaining process quality in almost all manufacturing processes in their design stage and throughout their lifetime.

In summary, from the engineering through commercial production stages of any kind of processes, process quality assurance means the clear identification of all quality conditions in every piece of equipment or process, and the establishment and maintenance of quality conditions which meet the five criteria for quality assurance.

• **EFFECTS**

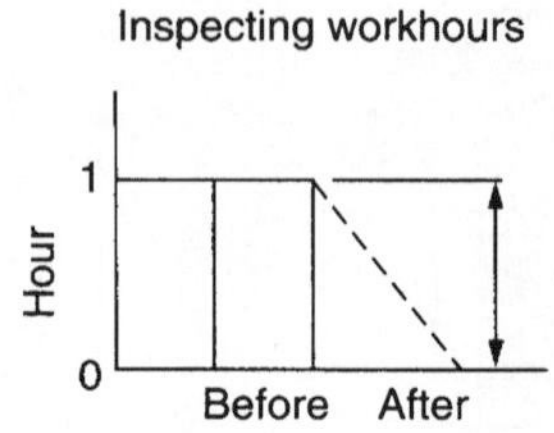

• **REMEDIABLE ACTIONS**

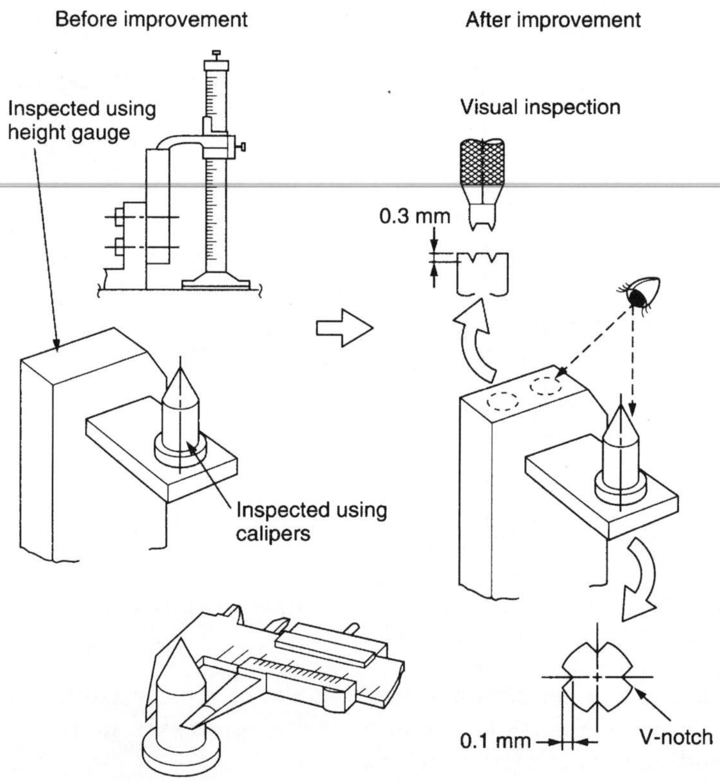

Figure 10.6 Changes in quality conditions are easy to detect.

10.3.6 An approach to Zero Defects

Thorough consideration of an approach to Zero Defects arrives at the conclusion that the occurrence of quality defects is prevented in each process. If this goal is not attainable, the outflow of defective products to downstream processes is checked by careful quality inspection at the exit of each process. There are the following three types of approaches:

1. Create a highly reliable process, the quality conditions of which totally meet the five criteria for quality assurance. Insofar as quality conditions are consistently monitored and properly corrected, defective products will never be manufactured.
2. Due to current technical levels or budgetary limitations, the five criteria for quality assurance are not totally satisfied. To compensate for this, outflow of defective products to downstream processes can be completely prevented by applying 100 percent inspection using automated equipment at the exit of each process.
3. Due to these same restrictions as noted above, neither total satisfaction of the five criteria for quality assurance nor 100 percent inspection by automated equipment is attained. Operators are, therefore, compelled to prevent manually the outflow of defective products to downstream processes.

The above approaches are divided into three steps, and are actually developed in the following sequence:

Step 6-1: Develop activities to prevent the outflow of defective products to downstream processes (remedies focused on quality results).

Step 6-2: Develop activities to prevent manufacturing of defective products (remedies focused on quality causes).

Step 6-3: Develop activities to maintain consistently quality conditions which have been achieved in the preceding steps to eventually attain Zero Defects.

The above steps are divided into more detailed substeps. They differ from one plant or process to another according to the extent of automation therein. Table 10.1 presents an example of operators' activities in a plant where repetitive manual work persists. These activities which strive after matters of process quality assurance must be developed in all quality-related departments, such as plant engineering, product design, production, maintenance, quality assurance, and so on.

Among the many and various efforts involved, the production department is in charge of relatively easy tasks. There is, however, no difference in the details of activities aimed at Zero Defects and Zero

TABLE 10.1 Dividing Step 6 into Substeps

Substep	Major activity
1	Remedies focused on quality results
1-1	Delineation of quality assurance flow diagram
1-2	Process quality assessment
1-3	Preventive measures against defective product outflow
1-4	Product handling
1-5	Can quality defects be detected when they occur?
2	Remedies focused on quality causes
2-1	Raw material control
2-2	Measuring apparatus control
2-3	Jig and die control
2-4	Machining condition control
2-5	Mistake-proof control
3	Establishment of process quality assurance system

Breakdowns continued from Step 1. In regard to matters of quality, operators apply identification tags to defective parts and difficult work areas discovered in equipment, and pose questions. They also write these problems into the four lists, and then cross them out when solved.

10.4 A Prerequisite for Process Quality Assurance

In the previous sections of this chapter, the discussion was mainly relevant to automated processes. There are, however, not many fully automated plants. On most of the shopfloors, operators actually manipulate operation consoles, set workpieces onto machine by hand, or make changeovers. Operators' roles and tasks are breifly reviewed in the subsequent sections. Further detailed discussions of and case studies in manual work are described in Chap. 12.

10.4.1 Operators' roles

The following are traditionally considered typical operators' roles in many companies:

- Manual operation and manipulation of equipment
- Cleaning and minor servicing of equipment, jigs, tools, and dies
- Transportation and control of raw material, component parts, workpieces, and finished products

- Disposal of dust, chips, scraps, and defective products
- Early detection of abnormalities in terms of safety, quality, and equipment conditions along with accurate and prompt corrective actions or reporting to the related departments
- Record-keeping on performance of production and quality
- Control of any other material as needed

Taking into consideration the past occurrences of losses, managers must review the operators' roles as follows:

- What are the operators' roles?
- Have operators sufficiently performed their roles?
- If not, what is the reason?
- What kind of work standards for operators are needed to perform their roles?
- What kind of skills do operators require?
- Do operators have required skills?
- If not, what can managers do to assist operators?

10.4.2 Materials to be handled by operators

Operators must handle numerous and diverse materials involving equipment:

- Jigs, dies, tools, measuring apparatus, auxiliary equipment
- Raw materials, component parts, workpieces, products
- Cleaning appliances
- Defective products, scraps, chips, packing wastes
- Pallets, containers, boxes, shelves, work benches
- Material handling equipment

Sufficient quantities of materials as listed above which meet the needs for production at all times must be provided at adequate locations. To facilitate this procedure, the following actions may be taken:

- Specify how much materials are to be used by whom, when, and where.
- Confirm whether the materials are functioning properly and are satisfactorily being maintained in terms of quality and quantity.
- In order to prevent needless searching, place materials where they can be identified at a glance; also what and how much is present.

- In accordance with frequency of use, specify locations and methods to pick up materials with the least labor.
- Decide on procedures for supply and disposal of materials, and also designate personnel in charge of the routine control of this function.

10.4.3 What is proper operation?

It is quite natural for operators to operate equipment properly. However, when a thorough consideration is made of "What is proper operation?," numerous kinds of problems are found to be hidden and persistent.

Although operator's work standards and procedures are standardized and compiled into manuals, most of these manuals may have been prepared by staff in the plant engineering or production departments, and then forced upon operators to follow in a traditional, one-way style.

No matter how attractively manuals are displayed on shelves in shopfloor offices, the actual critical operation, setup, and adjustment of equipment should rely on the experience and the sixth sense of veteran operators. A solution must be found for this situation where the people who set the rules are functionally separated from those who follow the rules.

As manual work is replaced with automated equipment, operators' work becomes easier and simpler, but the supporting control systems become progressively more complicated. Once trouble occurs in these automated systems, troubleshooting is very difficult. Except for a few specialists, control circuits and computer software become a kind of black box in many cases.

Furthermore, a misoperation in the automated processes results in incredibly large and serious losses compared with those in which a limited number of machine tools are operated manually along a conveyor. The production department managers in their routine supervision must, therefore, provide operators with new and adequate instructions as described here:

- The reason for operating in accordance with the given quality-related standards within the context of the structure, motion, and function of equipment
- How process and equipment constitute a system
- What kinds of influences stemming from a breakdown of component parts impact a piece of equipment or an entire system
- How equipment is operated properly

In reality, frontline personnel, including engineers, cannot keep up with the rapid progress made in automation because they do not receive sufficient education. Thus, human behavior causes most of the losses which result in greater damage to plant operations.

10.4.4 Learn quality control methods

During the preceding activities associated with the managers' models, managers must assess the knowledge and skills necessary for operators, as described in Table 3.1, to be responsible for qualitative matters; for example:

- The five criteria for quality assurance
- Inspection items in terms of quality results and quality conditions
- Problem solving methods such as where-where, why-why, and PM analyses and their applications to identify and analyze quality conditions

By way of stepwise education, managers and engineers then teach operators about each piece of equipment or process as follows:

- Relationship between the process quality and quality conditions
- Summary of remedies focused on quality results (Step 6-1) and quality causes (Step 6-2)
- Unresolved problems in Step 6-1 to be addressed as the most critical issue in Step 6-2

In addition to the above, if operators have not mastered enough of the basic concepts and methods of quality control in the previous steps, the weak points must be reinforced as needed, for example, a procedure for data collection, a Pareto chart, a histogram, quality control charts, and so on. It is also expected that required teaching materials were prepared in advance and operators were trained in Step 6-1 or the preceding Step 5 whenever possible.

10.5 Step 6-1: Remedies Focused on Quality Results

10.5.1 How to proceed with remedial actions

Operators made a large number of improvements to eliminate the breakdown of equipment in the previous activities through Step 5. Remedial actions in the matters of quality, however, are much more difficult than those of breakdowns. Operators, therefore, develop activities that target quality results prior to their consideration of quality causes. In other words, preventive measures against the outflow of defective products to downstream processes are taken first.

If operators attempt to deal with quality causes before they acquire sufficient technical knowledge and skill related to quality, it is absolutely impossible for them to succeed. These most difficult quality issues, among various other aspects of production technologies, require a steady step-by-step strategy.

Highly reliable equipment that manufactures no defective products by fully satisfying the five criteria for quality assurance, or, automated inspection equipment that prevents completely the outflow of defective products by 100 percent inspection provides an ideal condition in which no remedial actions focused on quality results are required.

In most cases, this kind of ideal state is not realized easily and immediately. Present technical know-how cannot eliminate the need for persistent efforts by operators to prevent manually the outflow of defective products. The following criteria, therefore, are introduced in order to evaluate human performance in this regard. The questions which relate to and refer to any quality-related objectives in terms of either manufacturing equipment or manufactured products must be assessed by way of the following procedure.

From operators' standpoint, are the existing rules:

1. *Clearly defined?*
2. *Well understood?*
3. *Well observeed?*
4. *Easy to observe?*
5. Are deviations *detected at a glance* if quality defects occur?

Although these criteria: "Clearly defined?, Well understood?, Well observed?, Easy to observe?, Be detected at a glance?" are applicable to various aspects of human performance and motion, they are utilized, in this case, for the prevention of outflow of defective products by adhering to the given quality-related rules. Therefore, they are called "the five criteria for ease of observation."

To conclude, the remedial actions focused on quality results to be taken in Step 6-1 are the clear identification of all qualitative factors to be incorporated into each piece of equipment along with the actual modifications of equipment and work methods which satisfy the five criteria for ease of observation, as diagrammed in Fig. 10.7.

10.5.2 Step 6-1-1: Prepare a quality assurance flow diagram

To begin with, it is necessary to identify and specify the details of process quality at the exit point of each piece of equipment, in order to prevent the outflow of defective products. Process quality is referred to as single or multiple elements of quality created and assured at each piece of equipment or process in response to product design. A chart for this purpose must be constructed by delineating equipment and its process quality from the entrance to the exit of the production line. This chart is named the Quality Assurance Flow Diagram (QAFD).

QAFD is drawn as illustrated in Fig. 10.8. It includes process quality (Qn) conceptually presented in Table 10.2 at the exit point of nth piece of equipment (En) along with its tolerance and designation of products to be inspected among the initial, interim, and final product in a certain unit of production as defined in Table 10.3.

In some extreme cases, the plant engineering and quality assurance departments cannot identify process quality for every piece of equipment, even in the commercial production stage. To avoid this kind of situation, the related departments must address themselves to matters of quality as one of the most critical issues from the earliest stage of TPM. Teaching materials for themselves and the operators, therefore, should be prepared as early as possible.

In some processes, quality inspection is carried out by the quality assurance department, and, as a result, operators are not concerned with that function. In these cases, when stepwise education takes place, engineers must teach operators the classification of quality (on-specification) products and defective (off-specification) products, as defined in Table 10.4, plus the basic concepts of quality assurance and data handling. It is expected that operators who were trained in the basic techniques of data collection and analysis in the preceding Step 5 are already applying these skills to routine aspects of their job.

These inspecting items, to be confirmed in terms of process quality in response to the specification of products and quality control standards, must be clearly designated and classified according to the following sequence:

1. Whether quality conditions of manufacturing equipment (quality causes) or quality of manufactured product (quality results) must be inspected.

2. If quality results need to be inspected, indicate whether it is to be the quality of the initial, interim, or final products.

After completing the above classification, operators make an assessment in accordance with two out of the five criteria for ease of observation (Clearly defined? Well understood?) as follows:

- Are the inspection items definitely needed? Are they clearly defined in the specified quality control standards?

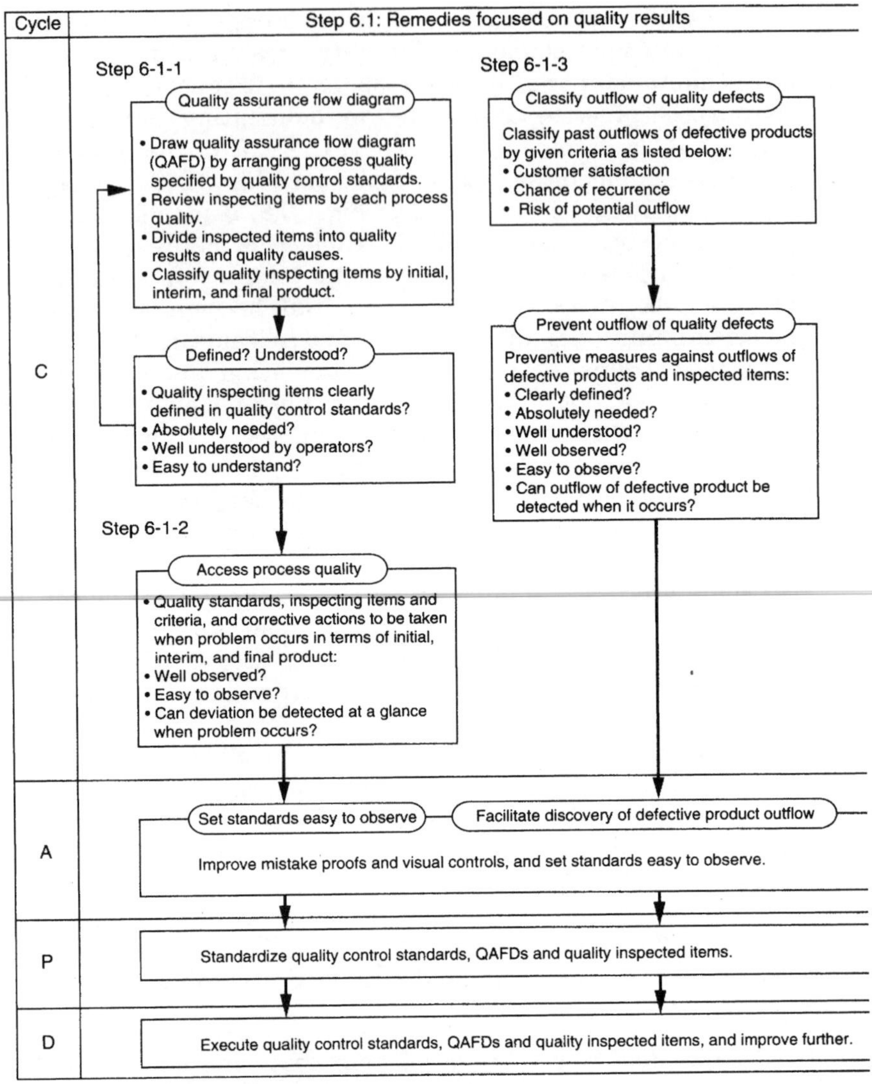

Figure 10.7 Step 6-1: Remedies focused on quality results.

- Do operators understand well what the inspecting items are? Are the specified standards easy to understand?

If errors or inadequacies are found in inspecting items and quality control standards, necessary corrections must be made. Of course, all operators need to be able to distinguish between quality and defective products defined in Table 10.4 as prerequisites in dealing with matters of quality.

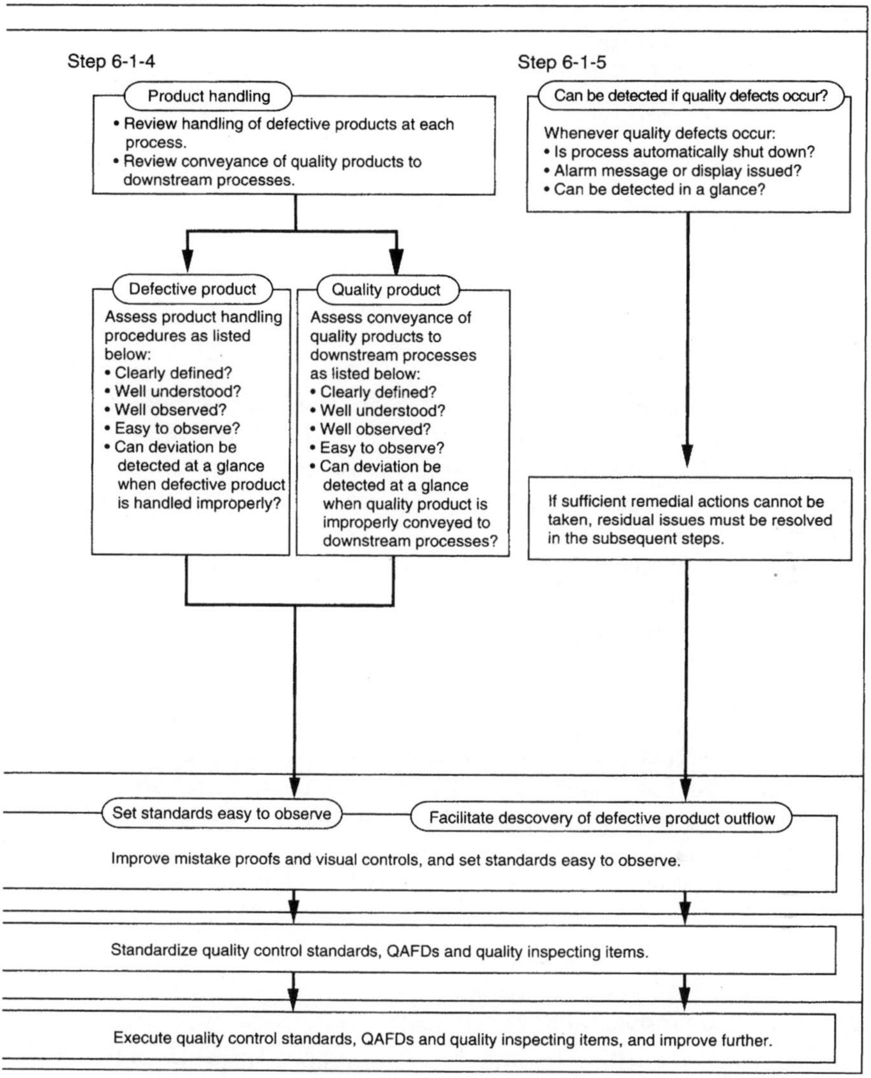

Figure 10.7 *(Continued)*

10.5.3 Step 6-1-2: Assess process quality

In the preceding Step 6-1-1, process quality and relevant quality control standards are assessed as to whether they are clearly defined and well understood by operators for each piece of equipment. In this present step, these standards and inspecting items are assessed according to the remaining three criteria ("Well observed?, Easy to observe?, Be detected at a glance?") in the order of initial, interim, and final products as follows:

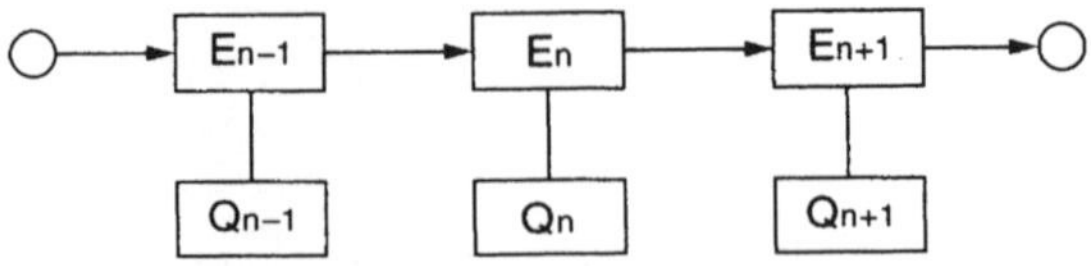

Figure 10.8 A Quality Assurance Flow Diagram (QAFD).

- Are the quality control standards faithfully observed by operators?
- Are the quality control standards easy to observe?
- Is deviation detected at a glance in the event of operator's negligence of quality control standards?

In this way, the question of whether each operator is capable of inspecting equipment along with process quality of manufactured products according to the quality control standards can be answered. Inadequate inspection methods or equipment must be corrected by applying visual controls and mistake proofs. As a result, any abnormal event, such as the occurrence of quality defects, operators' negligence in reference to quality-related rules, or inadequate operating conditions of equipment, is immediately detected by anyone. Of course, operators must be trained to be able to detect and remedy these undesirable situations, and to relay quickly accurate information to other concerned departments.

The above activities are repeated until the five criteria for ease of observation are fully satisfied. Any information regarding the outcome of these same activities should be recorded in quality assurance flow diagrams, quality control standards and any other relevant documents, especially when work methods or equipment are modified.

10.5.4 Step 6-1-3: Take countermeasures to prevent defective product outflow

Past occurrences of outflow of defective products are reviewed and classified in accordance with prescribed evaluation criteria. For this

TABLE 10.2 Process Quality (Qn) (Partial)

Quality	Tolerance	Initial product	Interim product	Final product
Length	±0.3	❍	---	❍
Width	±0.3	❍	---	❍
Thickness	±0.1	❍	---	❍

TABLE 10.3 Definition of Initial, Interim and Final Product

	Product/workpiece	Definition
Initial product	1. Commencement of work	Start of work at shift change or after break
	2. Changeover	At resetting of operating conditions to change raw material, or product/workpiece ▪ Raw material change ▪ Jig, tool or die change ▪ Operating condition change
	3. Adjustment	At minor adjustment under same operating conditions
	4. Jig or tool change	Due to interval for specified service, wear or breakage
	5. Alternation	Operation alternated by another operator
	6. Repair or service	At restart of operation after repair of breakdowns or periodical service
	7. Lot change	At change of product/workpiece lot
Interim product		Interim product manufactured using same raw material under same operating conditions
Final product		Final product manufactured using same raw material under same operating conditions

purpose, necessary data covering a prior period for a minimum of one year, along with possibilities of occurrence in the future, are investigated so necessary preventive measures against the recurrence of these same outflows are taken.

To make this possible, data collection schedules of quality defects are compiled when detailed TPM schedules are planned. Meanwhile, the quality assurance department must specify data collection and recording procedures by which operators keep the necessary data during their routine job. At least the following basic data must be recorded:

- Process where defective product has been or may be assumed to have been manufactured
- Process where defective product outflow was actually detected
- Any materials such as operating data, photographs, sketches, and actual defective products that may help in the forthcoming study of the type and details of defects and contributing operating conditions

TABLE 10.4 Definition of Quality (On-specification) Product and Defective (Off-specification) Product

Classification	Definition	
Quality product	Product or workpiece meets given quality control standards.	
Defective product	Product or workpiece does not meet given quality control standards. Store in colored containers and handle as specified below:	
Scrap	After evaluation, defective product must be scrapped.	Red
Rework	After evaluation, defective product can be repaired in current or upstream process.	Yellow
Hold	Evaluation is requested from quality assurance department.	Blue
Recycle	After evaluation, partial or whole product can be reused.	Green
Dropped parts	Parts dropped onto floor during processing, transporting, or assembling.	White

- Troubles in downstream processes caused by specific outflow of defective products
- Levels of difficulty in detecting the occurrence of defective product

By reviewing the results of a study based on the above materials, the implication of recurrences is assessed according to major indexes, such as customer satisfaction, frequency of occurrence, possibility of outflow, damage and troubles in downstream processes, and so on. The method of assessment, indexes, and criteria vary widely in different types of industry and plant operating conditions. Several examples are examined here:

Customer satisfaction

- Critical dissatisfaction: Defective products should be recalled lest the manufacturer should be prosecuted under provisions of a Product Liability Law.
- Dissatisfaction: Defective products may be recalled and replaced on a case-by-case basis.
- Complaint: Image of brand or company may be damaged.
- Full satisfaction: Defective products are not shipped out from factory or are not discovered without a specialist's careful inspection.

Frequency of occurrence

- A large number of quality defects occurred in the past with very high risk of recurrence in the future.
- A moderate number of quality defects occurred with high risk of recurrence.

- No quality defects occurred in the past, but the potential of occurrence exists.
- No quality defects occurred in the past; there is a very slight possibility of occurrence in the future.

Possibility of outflow

- No barrier at the exit of process. Off-specification products definitely flow out if quality defects occur.
- Visual or sampling inspection is carried out at the exits of processes with a high risk of outflow.
- One hundred percent inspection is carried out at the exits of processes by automated inspection equipment with very low risk of outflow.
- High reliability of equipment achieved with no possibility of outflow as long as quality conditions are properly maintained.

After quality defects are classified as critical, serious, moderate, and slight in accordance with the above procedures, the required remedial actions are decided upon while taking present technical levels and budgetary limitations into account.

Sometimes, necessary quality data are not kept due to ambiguous classification of process quality, or misunderstanding of the process in which defective products are manufactured. In these cases, additional and more detailed surveys must be conducted. When sufficient data is not collected by the starting point of the above procedures, those which were collected actively during the earlier stages of Step 6 must be used.

To prevent recurrence of the same quality defects, operators must correct equipment as well as their work methods. By making references to all of these five criteria, they then implement suitable new measures and prepare quality control standards by themselves on a trial basis, as listed here:

- Are the inspecting items and their criteria clearly specified? Are they actually needed?
- Are these same items and criteria well understood by operators?
- Are the quality control standards faithfully observed by operators?
- Are the quality control standards easy to observe?
- Can deviations be detected by anyone at a glance in the event of an operator's neglect of quality control standards or an occurrence of an abnormal state?

The above activities must be repeated until the five criteria for ease of observation are fully satisfied in a way similar to that conducted in previous steps. If effective countermeasures are not taken, residual problems must be pursued as further issues.

10.5.5 Step 6-1-4: Products handling

In the past steps of an autonomous maintenance program, various remedial actions were employed to correct problems in the handling of materials. In Step 6-1-4, the methods for handling products are again totally assessed in view of the five criteria for ease of observation. Initially, operators review the handling of defective products, such as location and methods of storage, work methods, and procedures, based on these same five criteria from the viewpoint of process quality assurance.

They then consider the conveyance of quality products to downstream processes in an approach similar to the one above. Because there are various types of conveyances, for example, hands, four wheelers, conveyors, motor driven vehicles and automated carriers, their modification sometimes has certain effects on overall material handling procedures in the entire plant, as well as computer-oriented process control systems. Therefore, thorough consultation and cooperation with the other related departments is essential. The key aim here is to prevent confusion which impacts downstream processes.

Based on the results of these assessments, operators must thoroughly improve work methods, tools, equipment, and jigs by applying visual controls and mistake proofs until the five criteria for ease of observation are fully satisfied.

At the end of this step, pertinent product handling standards and procedures must be revised. When materials are handled manually, related documents must be written clearly by specifying the "5W's and 1H."

10.5.6 Step 6-1-5: Are quality defects detected when they occur?

To complete remedies focused on quality results, operators check whether quality defects are detected whenever they occur in any process: If quality defects occur

- Does the process or piece of equipment shut down automatically?
- Is an alarm lamp or warning sound issued?
- Can anyone detect an occurrence of defective products immediately whenever they are manufactured?

Based on the results of the above assessment, equipment and work methods are modified thoroughly by applying mistake proofs, alarm lamps, warning sounds, and any other visual controls. If operators have not solved all problems, residual difficulties must be definitely pursued as outstanding issues in the forthcoming step.

10.6 Step 6-2: Remedies Focused on Quality Causes

10.6.1 How to proceed with remedial actions

Remedies focused on quality causes, that is, a series of activities to prevent the occurrence of quality defects, are developed by way of clear identification, accurate setting, and prompt restoration of quality conditions whenever they deteriorate. If, by this faithful maintenance of quality conditions, they are properly managed at all times, quality results created in a certain process should coincide with given quality specifications without difficulty and, as a result, off-specification products are never manufactured.

The general activities common to and developed by all departments which are concerned with matters of quality must be considered. The details to be dealt with by operators will be discussed in the following sections as needed.

To begin with, those quality-related materials which were provided to the production department must be thoroughly examined from the viewpoint of process quality assurance. Some pertinent factors that affect quality results are equipment, jigs, and tools involving quality control standards, work procedures, visual controls and mistake proofs.

In the previous steps, frontline personnel aimed at quality results, or, in other words, quality already produced. In Step 6-2, however, they make a total assessment of the relationships among quality specifications, quality causes, and quality results by taking suitable remedial actions against quality causes.

It should be recognized that almost all kinds of quality control standards or operation manuals presently offered by the plant engineering department or vendors describe very little in regard to the strict quality conditions which are addressed from now on. When something of a rigorous nature is written there, it is usually is neither clear nor helpful. All personnel concerned, therefore, must critically check the existing quality-related materials mentioned above.

In order to prevent the manufacturing of the same defective products, personnel need, by using the techniques mastered through the past TPM activities, to clearly identify and accurately assess the quality conditions surrounding each process in which quality defects occurred during the past year.

These techniques, namely, where-where, why-why, and PM analyses can be applied in making detailed observations and analyses of operating conditions in existing equipment so as also to identify favorable quality conditions. When these quality conditions are once definitely identified and assessed in light of the five criteria for quality assurance, necessary remedial actions can be taken in reference to tools, equipment, and jigs. Thereafter, suitable visual controls and mistake proofs can be applied.

Operators are likely to resolve considerably difficult problems when quality conditions are of a mechanical and visible nature, as indicated by some case studies described earlier in this book. They cannot, however, handle the problems in chemical, instrumental, or computer areas. Such problems are frequently difficult to solve even for experienced engineers. It is, therefore, key in seeking to reach targeted objectives, to allocate feasible assignments to operators and to coordinate their efforts with those of the plant engineering, product design, quality assurance, production, and maintenance departments.

10.6.2 The sequence of countermeasures

From the viewpoint of process quality assurance, quality conditions must be provided by equipment rather than unreliable human efforts and are addressed in the following sequence:

1. Identify quality conditions to be incorporated into a piece of equipment.
2. Assess quality conditions.
3. Remedy quality conditions.
4. Review and standardize inspecting items in terms of quality conditions.

These procedures are described in detail below:

1. *Identify quality conditions.*
   In Step 6-1, remedies focused on quality results are begun by the clear identification of process quality and the creation of its flow diagram. In Step 6-2, remedies focused on quality causes begin with the identification and division of very accurate and detailed quality conditions which meet quality specifications as diagrammed in Fig. 10.9.

   This initial preparation generally is carried out by engineers of the product design, plant engineering, and quality assurance departments. This kind of activity helps the engineers to acquire the necessary engineering skill mentioned earlier in this chapter.

2. *Assess quality conditions.*
   The quality conditions strictly identified through the above approaches are assessed as to whether they meet the five criteria for quality assurance, as shown in Fig. 10.10. When any of these condi-

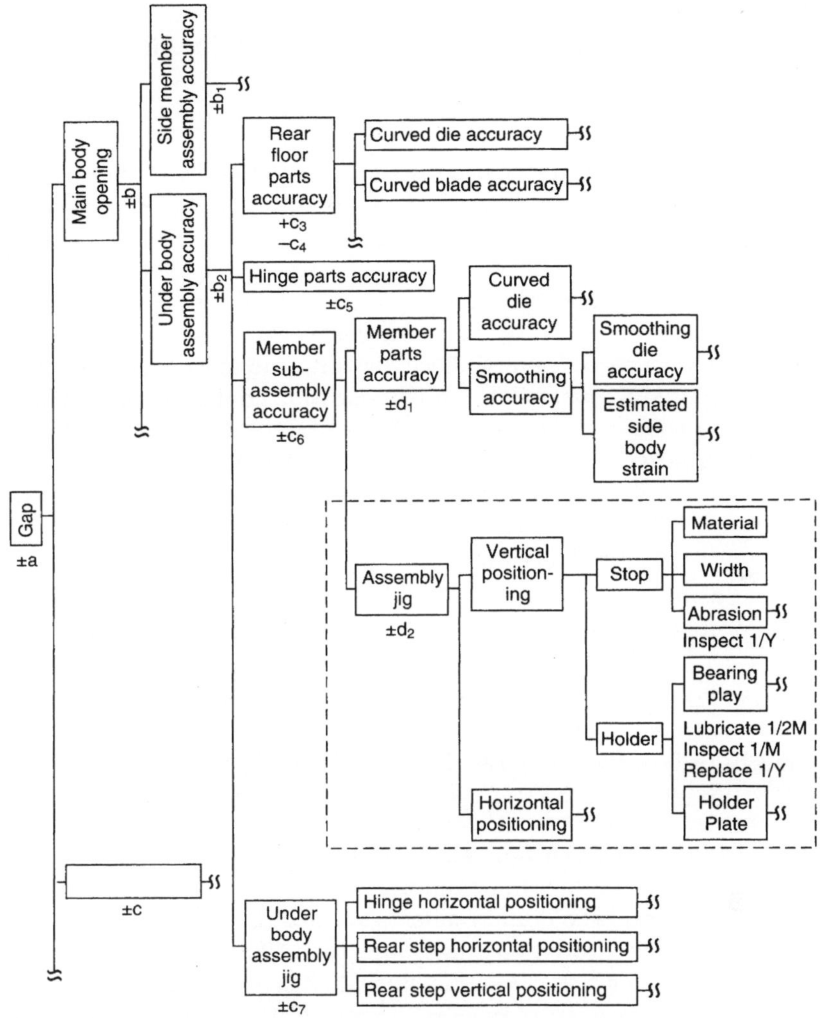

Figure 10.9 Identifying detailed quality conditions. [Note: Intervals are abbreviated using "frequency/initial letter" in this book. For example, 1/D (daily), 1/2W (biweekly), 2/M (two times per month), 1/Y (yearly), and so on.]

tions do not meet these same criteria, all corresponding work methods, tools, equipment, and jigs must be modified until these criteria are fully satisfied.

A considerable number of trial-and-error attempts may be needed to make a rigorous analysis of the quality conditions as described in some case studies described in Chap. 6. Therefore, very detailed observations on machine motion along with the occurrence of quality defects must be patiently and repeatedly conducted. To achieve significant results of improvement, the effective use of problem solving methods learned through previous TPM activities is essential.

At this stage of TPM, the recurrence of similar quality defects becomes very rare. Nevertheless, definite countermeasures must be taken against any suspect quality conditions detected in Step 6-1. In the event that there are some difficulties with the strategy to relate certain quality conditions to the recurrence of quality defects, further analysis must be conducted based on a technical, logical approach. These methods are the domain of the plant engineering and maintenance departments.

3. *Remedy quality conditions.*

The problems discovered during the assessment of quality conditions are resolved by improving related equipment, jigs, and tools, and applying more detailed visual controls or mistake proofs in order to satisfy the five criteria for quality assurance. In this way, operators and maintenance personnel easily inspect quality conditions during their routine jobs. Thorough implementation of visual controls makes it easier to detect any deterioration of quality conditions. If the deteriorated equipment is immediately restored, no defective product should be manufactured.

In summary, Zero quality defects are attained by the realization of highly reliable equipment, jigs, and tools. An example of remedial actions taken for a car assembly jig is described in Figs. 10.5 and 10.6.

In the long run, insofar as no abnormalities are found by simple, routine inspections, no particular corrective actions are necessary. The implication is, therefore, that it is possible to reduce adjusting time in start-up and changeover operation. In other words, Zero Losses in terms of setup and adjustment are also attained from the standpoint of the six big losses.

This kind of modification of equipment sometimes requires a large amount of time and expense. The decision as to whether to apply the results of the pilot project to all other similar equipment is based on an estimation of the benefit/cost ratio of the modification involved. In some cases, it might be concluded from a financial viewpoint that the occurrence of defects in existing equipment is tolerable if the outflow of defective products is absolutely prevented.

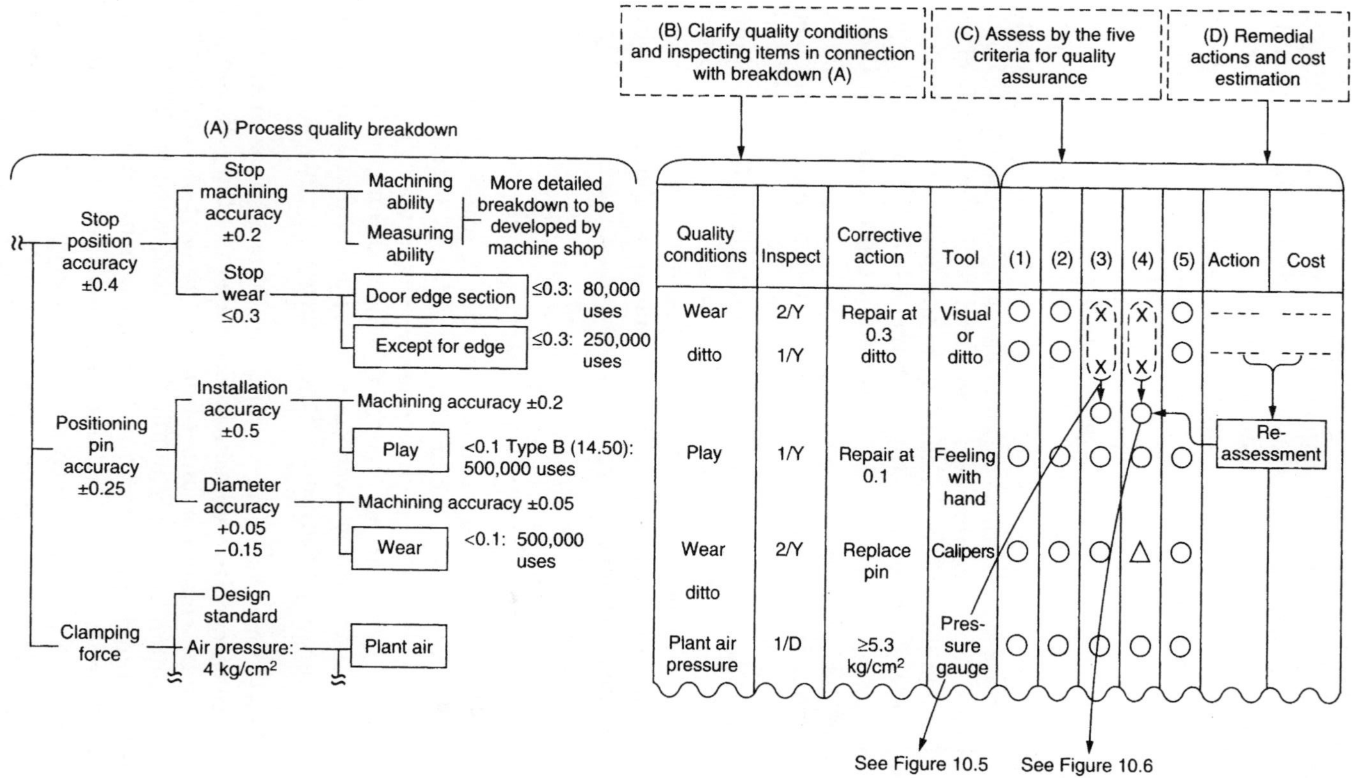

Figure 10.10 Assessing quality conditions.

Suppose, for example, a piece of equipment for which the defective quality rate once was 1:100 can be improved to a rate of 1:10,000 as the result of long and persistent endeavors in the previous TPM activities. It is, however, not so cost effective to decrease this rate to a one per million level if the timing of periodical shutdown maintenance or yearly turn-around can be carefully planned. The decision, whether the equipment is to be modified or not, must be made according to a managerial strategy in which the expenditure for the equipment over its lifetime is taken into account.

Apart from the question of whether modification of equipment actually takes place or not, the experience and knowledge acquired in connection with various actions through Step 6 must be incorporated into future equipment. Compared with the modification of existing equipment, design changes in the engineering or construction stages of a new plant are much more effective. Only this manner of approach can be called maintenance prevention in the true sense of the word.

All useful MP information must be compiled and standardized as engineering procedures in order for it to be effectively utilized by anyone at the factory or company level conducting effective planning and engineering for future construction and revamping projects. This kind of experiential information is, however, of very limited significance in many companies if it remains at the level of individual private know-how.

4. *Review and standardize inspecting items.*
   As the result of the efforts in Step 6-2, quality-related inspecting items determined and applied to some processes in Step 6-1 become unnecessary, and are replaced by the routine inspection and restoration of quality conditions. In addition, these same efforts simultaneously simplify other quality-related operators' tasks. All related inspecting items assessed, remedied, and set in Step 6-1 must, therefore, be reviewed again and standardized in light of the present operating conditions of each process.

The above four major procedures are repeated as many times as there are categorized subjects as described in Step 4. Those project teams which consist of personnel from the product design, plant engineering, maintenance, and quality assurance departments address more difficult and unresolved quality conditions in order to attain Zero Defects. Meanwhile, matters to be dealt with by operators are illustrated in Table 10.5, in which five categories such as raw material, measuring apparatus, jigs and dies, machining conditions, and mistake proofs are selected. Operators initiate their activities, beginning with the easiest task relating to raw materials, followed by most diffi-

TABLE 10.5 Dividing Step 6-2 (Remedies Focused on Quality Causes) into Substeps

	Substeps	Major activities
6-2-1	Raw material control	▪ Detail and classify raw materials. ▪ Totally inspect raw materials. ▪ Assess raw materials in terms of the five criteria for quality assurance. ▪ Set raw material control standards.
6-2-2	Measuring apparatus control	▪ Detail and classify measuring apparatus. ▪ Totally inspect measuring apparatus. ▪ Assess measuring apparatus in terms of the five criteria for quality assurance. ▪ Set measuring apparatus control standards.
6-2-3	Jig and die control	▪ Detail and classify jigs and dies. ▪ Totally inspect jigs and dies. ▪ Assess jigs and dies in terms of the five criteria for quality assurance. ▪ Set jig and die control standards.
6-2-4	Machining condition control	▪ Detail and classify machining conditions. ▪ Totally inspect machining conditions. ▪ Assess machining conditions in terms of the five criteria for quality assurance. ▪ Set machining conditions control standards.
6-2-5	Mistake proof control	▪ Detail and classify mistake proofs. ▪ Totally inspect mistake proofs. ▪ Assess mistake proofs in terms of the five criteria for quality assurance. ▪ Set mistake proof control standards.

cult machining conditions, and finishing with a review of mistake proofs. These major activities in Step 6-2 are summarized below.

10.6.3 Step 6-2-1: Raw material control

Operators review current control systems for raw materials in accordance with the five criteria for quality assurance. Based on these activities, raw material control is standardized.

10.6.4 Step 6-2-2: Measuring apparatus control

All measuring apparatus is listed to make a thorough assessment in context of these same five criteria. Any unsatisfactory apparatus must be remedied. At the end of this step, control standards for measuring apparatus are compiled in connection with the results of the assessment and remedies.

10.6.5 Step 6-2-3: Jig and die control

Regarding all jigs and dies, the same activities just described are repeated. Operators concentrate mainly on jigs. They accordingly review maintenance and inspection methods and criteria for jigs and dies. At the end of this step, control standards are set.

10.6.6 Step 6-2-4: Machining condition control

Having become adequately experienced with the five criteria for quality assurance, operators challenge the most difficult quality conditions of equipment. Because the plant engineering and maintenance departments carry out the same activities as illustrated in Table 10.6, allocation of assignments to the production department must be based on the extent of the operator's technical progress.

Operators achieve remarkable results when their objectives are confined to a narrow and limited area. The successful cases and experiences obtained through preceding managers' activities with pilot models are edited into one-point lessons for continuously teaching the operators little-by-little. Operators often achieve unexpected, successful improvement when they follow a proven model involving similar equipment.

At the end of this step, control standards for machine conditions are compiled. If remedial actions were taken by managers and engineers, the background and keypoints of these actions are taught to operators insofar as possible. Operators then prepare control standards. This approach helps to further the maintenance of machine operating conditions and enhancement of operators' technical skill in terms of quality.

10.6.7 Step 6-2-5: Mistake proof control

A large number of visual controls and mistake proofs are invented and applied in the past TPM activities. Moreover, many of those installed in Step 6-1 are no longer necessary if remedies focused on quality causes are successfully implemented and result in highly reliable processes. Operators, therefore, must list these aids and then determine which of them are essential when properly applied in accordance with the five criteria for quality assurance. At the end of this step, control standards for mistake proofs are prepared. Of course, the invention of new mistake proofs must be continued beyond this point.

TABLE 10.6 Quality Condition Control

Control items for quality conditions					The five criteria for quality assurance														
					Clear or quantitative			Easy to set			Resists variation			Easy to detect			Easy to restore		
No.	Items	Allowance	Interval	Inspecting method	Criteria preset	Criteria displayed	Easy-to-read criteria	Instrument installed	Easy inspection	Clear indication by instrument	Less wear	Less deterioration	Less deformation	Visible wear	Visible deterioration	Visible deformation	Easy relpacement	Easy adjustment	Restorative procedures documented
1	Variability in height of datum parts	0.01 mm	1/6M	Exclusive tool					Δ					Δ			Δ		
2	Tightening torque of datum parts	1.5 kgm	1/6M	Torque wrench									Δ			Δ			
3	Air pressure	2 kg/cm^2	1/M	Pressure gauge															
4	Alignment of clamps	±2.0 mm	1/6M	Standard workpiece															
11	Flow rate of coolant	250 l/min	1/M	Flowmeter						Δ									

No.	Detailed problems	Counter-measures
1	Change in wear	Periodical check with exclusive gauge
	Misalign-ment among 3 datum parts	Simul-taneous machining of 3 parts
2	Looseness in datum parts	Periodical check with torque wrench
6	Difficult to detect changes in pressure and flow rate	Installation of large warning lamp

Note: "Δ" indicates an unresolved problem.

TABLE 10.7

Step 6	Autonomous Maintenance Audit Sheets	Sheet 1 of 4			
No.	Audit points	Results			
1.	**Step 5 condition**				
1.1	Are all equipment-related problems solved? No residual issues?				
1.2	Is operators' routine maintenance definitely executed in accordance with autonomous maintenance standards and schedules?				
1.3	Are cause analysis and preventive measures taken in each event of breakdowns or minor stoppages occurring in spite of secure observation of autonomous maintenance standards? Assistance of full-time maintenance satisfactory?				
2.	**Group activity** (General)				
2.1	Are aims of Step 6 understood adequately?				
2.2	Is activity plan made in advance? Well executed?				
2.3	Are managers' models well understood?				
2.4	Is activity board adequately utilized?				
2.5	Are safety matters carefully respected?				
2.6	Are TPM activity hours and frequency adequate?				
2.7	Is more efficient way of TPM activity pursued?				
2.8	Are used spare parts and consumables recorded?				
2.9	Is meeting after on-site activity definitely held? Reports submitted?				
2.10	Is activity participated in by all members? No indication of dropout?				
2.11	Are all members cooperating equally? Not led by particular member?				
2.12	Are noteworthy ideas introduced actively to other PM groups?				
2.13	Is cooperation with full-time maintenance satisfactory?				
3.	**Substep activities**				
	• Use attached sheets.				
4.	**Equipment** (Main body and surroundings)				
4.1	Do status display and warning lamps function properly?				

TABLE 10.7 (*Continued*)

Step 6	Autonomous Maintenance Audit Sheets	Sheet 2 of 4			
4.2	Do safety devices function properly?				
5.	**Visual control**				
5.1	Are new visual controls devised? Noteworthy ideas actively revealed to other PM groups?				
6.	**Short remedial program**				
6.1	Is subject selected from the six big losses?				
6.2	Are problems clearly identified? Targets pinpointed?				
6.3	No easy countermeasures taken?				
6.4	Are cost and effects of program reviewed? Actual figures recorded?				
6.5	Are preventive measures against recurrence of problems provided?				

10.7 Step 6-3: Establish Process Quality Assurance

Autonomous maintenance programs, after a long and challenging term, approach their final goal at last. Starting with the most laborious initial cleaning, operators resolved diverse issues and problems. Those who engaged either in simple or complicated repetitive work, learned slowly how to identify problems in detail and take appropriate remedial actions.

In Step 6-3, as the actual final step of the autonomous maintenance program, operators totally review numerous and diverse standards prepared in the previous steps in light of the five criteria for ease of observation. It is most important that the results of these activities are precisely standardized and then rigorously applied on the shopfloor.

10.8 The Keypoints of an Autonomous Maintenance Audit

Activities in Step 6, as they are developed over several years, vary widely with plant configuration and operating conditions in each factory. An audit is, however, consistently carried out at the end of each substep. Because almost four years may have elapsed since TPM activities were launched, personnel or positions may have undergone considerable change.

TABLE 10.7 *(Continued)*

Step 6	Autonomous Maintenance Audit Sheets	Sheet 3 of 4			
3.	**Step 6-1-1**				
3.1	Are process quality and tolerances clearly identified at each exit of equipment? Quality assurance flow diagram (QAFD) drawn?				
3.2	Are process quality assurance lists prepared by reviewing quality-related modifications carried out in the past steps in terms of equipment, work methods, quality control standards, visual controls and mistake proofs?				
3.3	Are quality assurance flow diagrams revised whenever quality related modifications of equipment or work methods take place?				
3.4	Process quality and relevant inspecting items in terms of initial, interim and final product:				
(1)	Clearly defined? Absolutely needed?				
(2)	Well understood?				
(3)	Easy to understand?				
3.	**Step 6-1-2**				
3.1	Quality standards, inspecting items and criteria, and corrective actions to be taken when problems occur in terms of initial, interim and final product:				
(1)	Well observed?				
(2)	Easy to observe?				
(3)	Can deviation be detected at a glance when problems occur?				
3.2	Can everyone accurately and promptly inform relevant departments when quality defects occur?				
3.3	Are quality standards, inspecting item and criteria, corrective actions, visual controls and mistake proofs sufficiently improved?				
3.4	Are all quality-related documents revised after any kind of modification?				
3.	**Step 6-1-3**				
3.1	Are past outflows of defective products accurately classified in accordance with given criteria?				
3.2	Preventive measures against outflows of defective product and inspecting items:				
(1)	Clearly defined? Absolutely needed?				
(2)	Well understood?				

TABLE 10.7 *(Continued)*

Step 6	Autonomous Maintenance Audit Sheets	Sheet 4 of 4			
(3)	Well observed?				
(4)	Easy to observe?				
(5)	Can outflows of defective products be detected at a glance when they occur?				
3.3	Can everyone accurately and promptly inform relevant departments when defective products flow out?				
3.4	Are preventive measures against outflows of defective product and inspecting items sufficiently improved?				
3.5	Are all quality related documents revised after any kind of modification?				
3.	**Step 6-1-4**				
3.1	Are handling procedures for defective or quality products carefully assessed?				
3.2	Handling procedures for defective or quality products:				
(1)	Clearly defined?				
(2)	Well understood?				
(3)	Well observed?				
(4)	Easy to observe?				
(5)	Can deviation be detected at a glance when products are handled improperly?				
3.3	Can everyone accurately and promptly inform relevant departments when products are handled improperly?				
3.4	Are handling procedures for defective or quality products and inspected items sufficiently improved?				
3.5	Are all quality related documents revised after any kind of modification?				
3.	**Step 6-1-5**				
3.1	Is process automatically shut down whenever quality defects occur?				
3.2	Is alarm message or display issued whenever quality defects occur?				
3.3	Can everyone immediately detect when defective product is manufactured?				
3.4	Are mistake proofs, alarm displays and visual controls completely installed and improved further?				

Newcomers must be brought to understand major TPM concepts in the shortest period of time and join in the activities together with their experienced fellow operators. Those who have participated in autonomous maintenance activity since Step 1 can attempt to deal with the most difficult matters of quality. It is, to the contrary, unreasonable to have the same expectations of newcomers. Managers must, therefore, pay serious attention to this point and do their best to bring the technical competence of all operators to the same high level, insofar as possible. Otherwise, harmony in a group surely will diminish and some individuals or PM groups will drop out.

Those activities which address the matters of quality must be tackled by the production, maintenance, and plant engineering as well as any other related departments. This approach provides the best opportunity to bring technical skill to bear on the areas of product development and plant engineering. Anyone who participates in these activities learns very meaningful lessons by experiencing the actual occurrence of quality defects and the discovery of inadequate engineering performance in the past.

In Step 6, it is not easy, even for engineers, to identify quality conditions and make a detailed and thorough assessment according to the five criteria for quality assurance until they become familiar with kinetic, operating conditions of equipment, and practical matters on the shopfloor. Nevertheless, the mastery of engineering and production technologies in a true sense is not achieved without overcoming such difficulties.

The allocation of activities between autonomous maintenance and staff from other concerned departments must be carefully considered. The primary goal is to allow operators to proceed with their activities effectively by learning from the results of managers' models. An example of audit sheets is shown in Table 10.7.

Note

1. This case example and illustrations were adapted by the authors from Fumio Gotoh, Equipment Planning for TPM: Maintenance Prevention Design (Productivity Press, 1991), pp.134–144.

Chapter

11

Step 7: Autonomous Supervision

11.1 TPM Fully Established on the Shopfloor

In Step 7, knowledgeable operators conduct autonomous supervision and follow the standards set by themselves on the orderly shopfloor, where any deviation from normal or optimal operating conditions is detected at a glance by anyone. As expected, Zero Accidents, Zero Defects, and Zero Breakdowns were achieved. In other words, TPM is fully established on a company- or factory-wide basis.

Operators consistently maintain basic equipment conditions and restore deteriorated parts. On the other hand, maintenance personnel provide highly sophisticated planned maintenance by applying preventive maintenance to critical machinery.

Progress in the implementation of TPM differs according to the equipment or process configuration and its operating conditions, but at least five years are required to reach this status. If activities focused on qualitative matters are thoroughly carried out in Step 6 to attain Zero Defects, not less than another two years are usually spent addressing the most difficult quality conditions. It is quite a long journey to establish TPM.

There are, however, many companies who look at TPM from a narrow and short-range view when striving for autonomous maintenance activity, simply in order to increase productivity. In these companies, a partially established TPM system gradually collapses several years after given, limited targets are achieved. It is a matter of regret that group activities in this kind of circumstance are inactive because operators, along with leaders and frontline managers who feel self-satisfied, tend to be complacent.

Even though breakdowns and quality defects continue to decrease for several years after the accomplishment of the targeted TPM goals resulted from the initial efforts expended by all relevant departments, breakdowns and quality defects eventually increase as frontline personnel, including the plant manager, are replaced yearly and the endeavors spent establishing the TPM system are gradually neglected.

Unless autonomous supervision is firmly implemented up to Step 7 and based on a proper understanding of TPM concepts by way of sufficient repetition of the CAPD cycle, there is a large risk of deterioration of the TPM system, slowly, but persistently. All employees must, therefore, maintain the established TPM level and go forward, as illustrated in Fig. 11.1.

11.2 Maintaining the Current TPM Level

11.2.1 Maintaining activity

Operators and maintenance personnel, or, in other words, autonomous maintenance and full-time maintenance, cooperate with each other to maintain the TPM level established through the previous steps of the TPM development program. Operators inspect for the deterioration of equipment and then restore it in order to maintain, with certainty, basic equipment conditions. In the meanwhile, maintenance personnel lend their best efforts to enhance their maintenance skills for reaching higher technical levels. Of course, breakdowns in critical machinery also must be prevented by the pertinent monitoring of machine conditions, such as vibration.

As specified in routine and periodic maintenance schedules, operators and maintenance personnel must perform their assigned duties. In this way, optimal routine servicing intervals are established over several years on the basis of actual experience with the ongoing reduction of intervals allocated to cleaning, lubrication, inspection, condition monitoring, parts replacement, overhaul, and so on. All useful information obtained in these maintenance activities is fed back as MP information to the plant engineering department, so that it may be applied in the planning and engineering for future plant development.

For this purpose, it is an absolute prerequisite that operators definitely and continuously follow the standards set by themselves in the previous steps. In addition, validity dates of standards are clearly written into the standards. Of course, relevant standards and documents should ideally be revised whenever any changes or modifications take place in association with either hardware or software.

However, these revisions are, all too often, easily forgotten or neglected. Any essential written materials, therefore, are required to be revised whenever these designated validity dates expire. When this

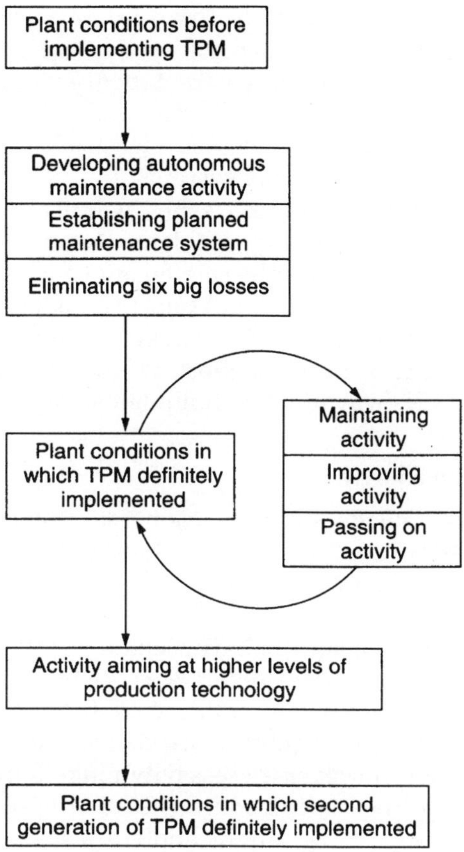

Figure 11.1 The development of TPM activity.

regimen of procedures is neglected, no one follows any rule that deviates, to any extent, from reality on the shopfloor.

11.2.2 Improving activity

Even though breakdowns and quality defects are reduced to 1:50, 1:100, or even further of the bench marks measured at the commencement of TPM implementation, Zero Breakdowns and Zero Defects, in the strict sense of the word, are never achieved yet. Certainly, entire factory conditions at any workplace might become unbelievably improved. Although Zero Breakdowns, as well as Zero Outflows of defective products from a factory on a monthly basis, are attainable, it is extremely difficult to maintain them throughout an entire year, despite the half decade of dedicated endeavors by all employees required to initially achieve them.

Existing equipment, because of financial restrictions, is not often discarded and replaced by new expensive equipment, in spite of the latter's high performance. Such cases are not rare in any factory. Upon a closer look throughout a plant, numerous attempts to remedy or automate are discovered in many processes, pieces of equipment, jigs, tools, work methods, and other support systems.

Operators strive to address these issues by repeating the CAPD cycle. This approach to remedy existing problems based on the Zero-oriented concept is an activity that must be carried out continuously as long as a company exists as a manufacturer. A short remedial program, pursued from Step 2 onward, is designed to make it possible for operators to maintain this kind of continuous effort. Of course, any useful information arrived at in this way must be compiled as MP information.

11.2.3 Passing on activity

In any company, employees are always being replaced. Personnel and positions are most often changed at the level of frontline managers. As a result, the great effort of all employees expended for the implementation and final establishment of the TPM system tends to be forgotten gradually. Eventually, people believe a clean shopfloor with Zero Accidents, Zero Defects, and Zero Breakdowns occurs quite spontaneously. This phenomenon brings on the collapse of the TPM system, which is followed by the return to original conditions of the shopfloor.

This fact does not, however, necessarily make apparent the need to repeat the same journey to introduce the TPM system to newcomers. They must, nevertheless, be taught and trained in regard to the fundamental TPM concept and methods in association with autonomous maintenance activities such as "cleaning is inspection," maintenance of the basic equipment conditions (cleaning, lubrication, and tightening), the four lists, activity boards, autonomous maintenance audits, operators' routine maintenance, short remedial programs, autonomous supervision, and so on.

These key methodologies to sustain TPM must, by careful planning, be systematically passed along to or inherited by all levels of an organization, such as the entire company, division, factory, department, team, and PM group. Some companies arrange for a TPM day or week to be held several times a year. On these occasions, all frontline personnel try to execute their designated work in accordance with relevant standards directed toward cleaning, lubrication, inspection, visual controls, mistake proofs, material handling, quality conditions, or other suitable topics.

Needless to say, on a factory-wide basis, all employees review simultaneously any rules to be followed in reference to the five criteria for ease of observation ("Clearly defined? Well understood? Well observed? Easy to observe? Be detected at a glance?"). This is the most convenient and reliable method to check on the degree of conformity to rules.

11.3 Aiming at a Higher Level of TPM

11.3.1 Two types of activity

After TPM goals are nearly achieved, what must be done next? The answer differs according to the conditions affecting operations and production, as well as the fundamental TPM policy in each company or factory. Generally, however, the answer is presented having the following two dimensions.

1. *Infrequent equipment updates*
   This type is typically found in the process industries, such as petroleum, chemicals, and cement. Each one of these processes produces a single type of product. These processes are highly developed in a technical sense and their products have a long commodity life and present little opportunity for innovation.

   In general, plants in these industries are, once constructed, operated as long as possible, that is, until intolerably high maintenance costs due to natural deterioration and aging do not allow further operation. Insofar as no incident such as an oil crisis occurs, large investments to modify these plants are very rare. At most, minor modifications are undertaken, for example, the upgrading of a bottleneck process to increase production output, the replacement of worn pieces of equipment, or the modernization of instrumentation along with the installation of more sophisticated computer-aided control systems.

   The previous TPM activities are sufficient to adequately operate the plant. There is almost no room to develop any further action. It is, however, not so easy to maintain a given level of TPM because such a plant is operated by a minimum number of operators and maintenance personnel. In addition, required maintenance costs, relatively high even during early days of operation, increase yearly due to natural deterioration. The most critical issue, therefore, is the reduction of maintenance costs.

2. *Frequent equipment updates*
   This type is found typically in assembly industries, such as automobiles, electronics, and other consumer goods. Rapid change in consumers' tastes and the variety of goods result in their short life as salable commodities. Drastic and frequent model changes along with innovations in supporting basic technologies take place contin-

uously. Although rapid automation is in progress, manual work still persists on the shopfloor.

In these industries, very tough competition continues in the worldwide marketplace. Those companies who delay in the development of new products or innovation, as well as those who make serious mistakes in sales strategy, are doomed to failure. It is essential for a company to launch new products which are welcomed by the consumer in a more timely way and are more reasonably priced when compared with those of its competitors.

11.3.2 Apply TPM experiences in future products and plant engineering

Never interpret TPM simply as activities that increase productivity or aim at short-term goals and benefits resulting from the elimination of the six big losses. In companies that do so, operators must be reintroduced to autonomous maintenance activities whenever the plant is revamped.

Ideally, a demanding program, such as TPM, implemented with a great effort on the part of all employees is not needed if easy-to-manufacture products are produced using easy-to-operate equipment that is properly maintained by operators and maintenance personnel knowledgeable in both equipment and quality on an orderly shopfloor at all times.

This utopian scenario, however, does not occur in reality. On the contrary, the essential matter that all manufacturers must deal with is survival in the face of rigorous competition, including mass-production technologies founded on product planning and development capabilities. Manufacturers, therefore, must put highly reliable processes on line more rapidly than their competitors.

Crucial sources of a company's competitiveness reside in the previous considerations. Because of this fact, there is no one to teach any manufacturer about mass-production technologies nor do any suitable textbooks exist in this regard. Necessary know-how must be obtained, on the other hand, through efforts expended by the company itself and all of its employees through various aspects of TPM concepts.

Attempts to remedy any kind of defects which exist in the present operations of a plant should primarily be carried out by engineers rather than by operators. These efforts also must be based on a thorough identification of the engineering weaknesses in existing equipment and products involved. Only through practice and experience involving the kinetic operating conditions of equipment installed in one's own plant and the resulting quality of products do frontline personnel learn essential mass-production technologies.

Management, on the contrary, frequently aims too exclusively at maximizing productivity rooted in a narrow and shortsighted vision, and then, ironically, boasts about the effective results of improvement. From another point of view, those improvements are recognized as the discovery and remedying of mistakes made by the product design and plant engineering departments in the past. Leaders who frequently mix their responsibility with that of shopfloor operators and maintenance technicians should be aware of these consequences.

Because products and equipment actually exist and are visible in the commercial production stage, anyone who makes a genuine effort attains effective improvement. However, during the engineering stage of a product or plant when only drawings or prototype models exist, it is not so easy to eliminate the causes of potential problems that are expected to occur in the forthcoming commercial production stage. This fact confirms the need to achieve a command of mass-production technologies.

Exaggerating somewhat in speaking about company conditions before the implementation of TPM, one may state that plant engineers understood equipment only in terms of drawings and vendors' catalogs, while maintenance personnel understood it only in regard to the replacment of broken parts. Furthermore, the operators, who were well aware of the kinetic operating conditions of equipment, possessed insufficient language skills for communicating their observations of problems accurately to staff. The product design personnel, on the other hand, understood nothing about production technology and were never interested in matters relevant to equipment. This was once the typical reality at the production frontline in many companies.

These situations, however, are eliminated by the arrival of Step 7. Moreover, the necessary groundwork is laid for meeting the challenge of reaching a higher level of mass-production technologies. Based on the experience acquired in the past TPM activities, operators and maintenance personnel participate by offering their ideas in a design review meeting during the engineering stage. By way of this kind of effort involving all employees, activities to assure process quality are developed in which easy-to-manufacture products and easy-to-operate and -maintain equipment are realized for the first time.

On the other hand, the product design and plant engineering departments actively share in various activities to eliminate the six big losses and participate in the autonomous maintenance audits. Sometimes engineers must assist operators and work together with them. These major efforts should be directed toward comprehending the reality of operating conditions and the weak points of their engineering capabilities.

11.3.3 What must be done next?

The second generation of TPM is not yet realized. For any company who arrives at Step 7, two major activities remain to be performed: automation and computerization. The basic aspects of automation are described in this book. The approach to computerization is examined briefly here.

Since computers became popular in the manufacturing business during the 1960s, companies made great strides in computer utilization. Computers were introduced in many relevant and easily applicable areas, such as finance, administration, sales, inventory, production control, and so on. Recently, many companies are implementing computerized facilities and maintenance management systems influenced by TPM concepts.

Daily troubles, however, do not consistently decrease, as discussed in Chap. 1, because most of these information systems are developed mainly by systems engineers. Their ideas, unfortunately, are founded on the mechanisms and functions of computers, rather than on the realities of operations and production on the shopfloor.

On the other hand, manufacturers' employees are still pressed by either simple or complicated repetitive work, while they are at all times handling either materials, equipment, or information. The first generation of TPM development focuses only on equipment and materials. Integrated activities in the second generation, therefore, need to address information rather than equipment and materials.

However, activities directed at information are often undertaken by the systems department which has no experiences in actual operations and production. To deal with this problem in many companies, systems engineers make a survey of actual conditions and interview relevant frontline personnel. On the basis of these preliminary studies, a computer system is developed and/or a suitable software package is purchased.

This style of approach is quite similar to the situation that is frequently referred to in this book in connection with the rules that are set by staff and then forced on shopfloor personnel. When input from the front line is neglected, meaningless information floods the company and managers are always struggling with computer terminals. Ironically, this kind of computer implementation results in additional, nonproductive, repetitive work.

With these observations in mind, it is obvious that the key to the introduction of the next generation of TPM consists mainly of all employees repeating the CAPD cycle from the viewpoint of information. All frontline departments from sales to production and distribution, and, of course, engineering, finance, or other supporting administrative departments, must undertake the overall assessment of information, as well as the matter of human work losses. Remedial actions against the source of losses and difficult work areas must be undertaken by following basic strategies which encourage frontline personnel. These same personnel then set their own standards by compiling the results of these remedies.

Utilizing the results of these activities, the systems department should devise suitable interfaces and protocols among departments and then provide computer systems to free human beings from numerous and diverse kinds of repetitive manual work. Human work thereby is shifted to more intellectual activities. If computers are utilized as one of the means to reach that goal, information handling work is automated. Productivity in supporting departments is also improved. Thereafter, troubles that previously flooded the shopfloor recede.

Nevertheless, providing these seemingly convenient systems does not always yield beneficial results. By leaving their development to a small number of specialists, these systems become a kind of black box as they expand enormously. Systematization by way of this kind of approach results in the fixing of structures and flows of information at a given point of time. In other words, it means that corporate organization and management are conservative and lose the flexibility required to adapt to quick and drastic changes in the worldwide marketplace and to the passage of time.

No matter how productivity is improved on a short- or mid-term basis, the consequences have no meaning if the most critical trait of organizational flexibility is lost. In this consideration lies the exact reason why the CAPD cycle aimed at information and human work must be repeated in an environment of total participation by all employees throughout the next generation of TPM.

This strategy assures optimal interfaces, protocols, and flows of information among relevant departments as well as individuals by means incorporating adequate redundancy and human interpretation into the system. Flexible corporate organization and management is thereby attained and human beings are freed from monotonous, repetitive manual work at long last.

Chapter

12

Autonomous Maintenance in Manual Work Departments

12.1 The Basic Concepts of Autonomous Maintenance

12.1.1 Autonomous maintenance is applicable to manual work

In Japanese assembly industries, repetitive manual work has been rapidly replaced with machinery since the 1970s when manufacturers faced a series of drastic problems caused by the oil crisis. Today's TPM concepts developed as one of the means of survival in a tough economic environment, and grew as a result of numerous and diverse trial-and-error activities conducted in many companies.

On the other hand, a large number of manual work shopfloors still exist where no remarkable machinery was introduced or only on a small scale limited by technical difficulties or prohibitive cost. Although TPM concepts originally were believed to be applicable mainly to matters of maintenance in plants relying on equipment, they are also relevant to improving PQCDSM within the context of repetitive manual work.

Autonomous maintenance activities in manual work-oriented departments are developed on the basis of concepts identical to "cleaning is inspection," steering organizations, overlapped small group organizations, repetition of the CAPD cycle, training of knowledgeable operators, realization of an orderly shopfloor, activity boards, and the four lists, such as equipment-oriented production departments. If a factory has both manual work-oriented and equipment-oriented production departments, two types of autonomous maintenance activities take place simultaneously. The difference between the two resides in the objectives of the activities; i.e., equipment and human performance, and the related composition of the TPM development program.

One might sense an inconsistency in referring to the operators' efforts to improve their manual work as "maintenance," in spite of the existence of little or no equipment in connection with their routine tasks. Because it is not so essential to distinguish between these same efforts made in equipment-oriented and manual work-oriented departments, operators' activities aiming at the realization of autonomous supervision on the manual work shopfloor are referred to as "autonomous maintenance" in both cases.

12.1.2 Is repetitive manual work fully improved?

Since F. W. Tailor late in the nineteenth century set out to enhance productivity, enormous and persistent efforts were expended based on industrial engineering concepts as one of the key issues on the shopfloor until the 1970s. On the other hand, Japanese quality control concepts were uniquely developed through standardization and documentation.

In spite of both kinds of efforts, operators' errors were not diminished on the manual work shopfloor as, for instance, in automobile assembly lines. Several factors that are conducive to an increase in the occurrence of assembly mistakes and scratching of body surfaces include wide fluctuations in demand due to seasonal and model changes, and the resulting frequent rearrangement in production cycle time, including the allocation of operators and process configurations. Furthermore, this trend became more serious yearly because of increasingly varied consumers' tastes and tougher competition in the worldwide marketplace.

A large volume of manual work losses (human errors + quality defects) still occur on the shopfloor that appear to be thoroughly improved according to established, traditional, managerial concepts. In reality, however, these losses are simply unrecognized due to the conventional way of thinking or neglected due to a lack of technical knowledge.

An attitude of resignation may be developing among manufacturers' management which believes that there is no more room for improvement in repetitive manual work other than by way of further automation, which is being implemented on a large scale. How, then, are autonomous maintenance activities effective even for manual work? Discovery of this reason resides in the discussion of what is key to the development of operators' TPM activities on the shopfloor.

12.1.3 The rules must be set by those who follow the rules

Even though no remarkable equipment is installed on the manual work shopfloor, decisions about the choice of hardware and software are made by frontline managers and engineers in the production and process engineering departments. These same personnel then attempt to force the operators, who had no voice in these regards, to use successfully the hardware such as process layouts, jigs, tools, measuring apparatus, parts shelves, containers, and vehicles along with software such as work standards and procedures, inspection criteria, display lamps, labels, and kanbans.

The manual work-oriented shopfloor has a much longer history than the equipment-oriented shopfloor and also experienced innumerable modifications of processes. This kind of history should allow for the development of much improved and more sophisticated process layouts and routine supervisions. Many problems, however, still remain and they are rooted in the situation, wherein those who set the rules are isolated from those who must follow the rules.

Frontline managers and engineers, who have no experience in actual repetitive manual work or possess only obsolete knowledge acquired a long time ago, decide upon the line layout, and then select tools, jigs, and measuring apparatus. They then prepare work standards and procedures in response to product design and specifications, and give them to operators.

In this style of management, the viewpoint of operators who must follow the rules is neglected or ignored. Consequently, questions such as these are frequently forgotten: "What is the safest and most convenient tool, jig, and measuring apparatus?," "Are existing process layouts, work sequences, and assembly methods adequate?," "Why do human errors occur?," "Can products be designed for easier assembly?" and so on.

Look, for example, at the procurement of four wheelers to transport materials in which, as an apparently minor issue, no one is generally interested. Some likely wheelers as listed in vendors' catalogues are selected and supplied, without pertinent considerations, to the operators. On the contrary, numerous experiments should be conducted beforehand by altering the structure of wheelers and taking into account various factors such as the strength and height of the bed, shape of the handle and position of wheels in relation to the dimensions, shapes, and styles of materials being transported. Only then are the safest and most convenient wheelers purchased or fabricated according to data obtained through these experiments. How many companies make such a careful analysis in connection with the various details of manual work?

Without a doubt, a thorough assessment of manual work is lacking. The reason for this neglect stems from the fact that a very limited number of staff personnel are absolutely unable to decide on a huge number of issues to be carefully considered. It is, however, even more unreasonable to permit these same personnel, who are strangers to actual manual work, to decide upon hardware and software. They then impose their decisions on the shopfloor. Surely, the ultimate mistreatment of operators is demanding of them, under these circumstances, to improve quality, increase production output, reduce cycle time, and eliminate assembly mistakes.

When the choice of hardware and software is decided on the basis of past customs and experience without careful consideration of current factors, operators never avoid either being forced or merely pretending to follow directions. It is just this situation that is responsible for manual work losses and industrial accidents not being eliminated in most factories, despite their long history.

The solution to this problem lies in operators deciding upon work standards and procedures, tools, jigs, and installation of parts shelves and containers. Rules set by those who must follow them are those that are easiest to implement with the least irregularities and waste. Human beings are not machinery and, therefore, undoubtedly make some mistakes. If, however, operators are allowed to respond with practicable rules they believe in, manual work losses are reduced to a great extent. Otherwise, accuracy and continuity of human motion, which are most essential in repetitive manual work, can never be realized.

12.2 How to Develop Autonomous Maintenance

12.2.1 Make original plans

The reality of repetitive manual work differs widely with operating conditions on each shopfloor. Frontline managers and engineers must prepare plans for a unique TPM program based on a thorough understanding of the concepts and methods of autonomous maintenance.

Accordingly, all manual work departments must do exclusive planning for autonomous maintenance activities to fit the characteristics of their own shopfloor. However, planning of TPM activities for a manual work-oriented department, which has no concrete target such as troublesome machinery, is much more difficult than for an equipment-oriented department.

Planners should not simply imitate the examples of other companies or in-house departments, but, instead, should request the creation of unique ideas according to experience attained with their own manager's model on a trial-and-error basis. This course of actions must involve thorough discussion among the various layers of managerial organizations of each factory. It is essential from early on to prepare carefully a plan associated with a critical examination of problems involved in current operating conditions and to arrive at a consensus on remedial actions to be taken.

12.2.2 Identify manual work losses (human errors and quality defects)

In repetitive manual work, TPM aims at improvements of PQCDSM. Accordingly, planning of autonomous maintenance activity must begin with a detailed assessment of present shopfloor conditions in order to identify concrete obstacles that stand in the way of achieving the desired objectives.

Although equipment-oriented departments encounter the six big losses as concrete obstacles, it is almost impossible to find analogous common losses for numerous kinds of manual work. There possibly may be some correlation between manual work losses and breakdown losses in equipment-oriented TPM. The expression, "manual work losses," however, has many meanings on the car assembly line, as for example, finishing mistakes using hand tools, improper bolt tightening, misassembly, and nonassembly of parts. In another example, the shipping department has, as manual work losses, mistakes in storage, packing methods and materials, delivery destination, and recording of packing numbers.

Therefore, a thorough assessment is needed to identify every impediment to the improvement of PQCDSM so as to define the specific losses on each shopfloor. Most importantly, all employees from factory leader to floor personnel should understand exactly what the losses are and then develop autonomous maintenance activities within the framework of a program specified for achieving the complete elimination of losses.

It is, however, very difficult to identify losses and recognize poor management in the past on a shopfloor that was operated in a conventional mode for a long time. If these same losses were adequately recognized by the established industrial engineering concepts, repetitive manual work would have been much more sophisticated long ago. If that were the case, there might be no need for implementing TPM now.

Because manual work has a long history and has undergone various attempts at improvement, too much emphasis may be placed on past experience and performance which often hinders, instead of helping in, a genuine search for innovative and breakthrough ideas. The most simple and reliable method to expose hidden losses is to make a comparison of present with ideal conditions: i.e., Zero Losses.

Therefore, the comparisons of the present status of a department with its past, one department with another, or one company with another are unproductive. Through the comparison of present operating conditions and problems with the absolute ideal, Zero, manual work losses as well as their circumstances and problems involved in the current shopfloors are, for the first time, thoroughly assessed without being restricted by past convictions and a conservative way of thinking.

It may, however, be impossible to eliminate completely losses resulting from human behavior. Nevertheless, remedies involving the application of visual controls, mistake proofs, alarm display, and so on should be taken in an effort to reduce this category of losses by way of Zero-oriented approaches.

12.2.3 Specify indexes to measure TPM progress

In the case of the six big losses in equipment-oriented departments, the highest levels of Japanese manufacturers cluster around 85 percent of overall equipment effectiveness. It is, however, difficult to find common indexes and goals for repetitive manual work to compare with other departments, factories, or companies because of the large variability in actual operating conditions.

Therefore, prior to the commencement of autonomous maintenance activities, major TPM issues and measuring indexes must be delineated to estimate the progress made in overcoming obstacles to PQCDSM. First of all, the status quo is set as the bench mark. Then targets and goals for each step in the seven step program of autonomous maintenance are decided upon with clear figures or percentages set for these established indexes. A combination of targets and schedules for achieving them constitutes the master plan for the TPM development program.

For example, major indexes in the car assembly line may include assembly line availability, quality rate [(cars produced – cars repaired) ÷ cars produced], labor for car assembly, quantity of misassembled and nonassembled parts, and so on. By an appropriate combination of these indexes, overall line effectiveness is calculated.

12.3 How to Construct the Seven-Step Program

12.3.1 Repeat the CAPD cycle

This book is written on an assumption that all autonomous maintenance activities are divided into a seven-step program. Although it is not necessary always to do so, as far as adequate repetition of the CAPD cycle is concerned, implementing these same seven steps uniformly within all departments makes it easier to develop TPM simultaneously on company-wide or factory-wide scales.

The program must begin with cleaning, the activity that is most easily participated in by any employee, like Step 1 in an equipment-oriented department's program. Operators then proceed gradually to more difficult activities by learning about work methods, the structure and function of their products, assembly tools, jigs, and measuring apparatus.

The key to the successful planning for the configuration of a seven step program is to provide adequate repetition of the CAPD cycle as many times as possible by dividing the procedure into three periods, as listed here:

- Period 1: Establish the basic conditions.
- Period 2: Establish the usage conditions of tools and jigs, work methods, and other materials to be dealt with.
- Period 3: Establish standardization and autonomous supervision.

At the end of each step, a substantial part of the expected results must be forthcoming. Planned goals are attained in the final period and consist of definitely mastering the CAPD cycle and acquiring the ability to perform autonomous supervision.

12.3.2 Period 1: Establish basic conditions

The basic conditions of manual work must be established similar to those in equipment-oriented departments (cleaning, lubricating, and tightening). In an assembly process, for example:

- Unnecessary materials do not exist on the shopfloor.
- Materials and methods for storage are clearly specified.
- All materials on the shopfloor are easily be recognized by anyone by using labels, tags, and other suitable aids.

Major activities to establish the above conditions are:

- Clean the shopfloor thoroughly.
- Make a thorough assessment of tools, jigs, and measuring apparatus in use. Locate them in designated areas if used or discard them if they are not used.
- Thoroughly remove unnecessary materials from the shopfloor.

- Assess dropped parts and take countermeasures, if needed.
- Eliminate the sources of contamination, such as wrapping paper, packing materials and chips, and dirt from parts containers and pallets.
- Clearly distinguish among products, workpieces in process, workpieces to be reworked, and quality and defective products.
- Review the shape, function, dimension, location, and method of installation for parts shelves and containers.

Following the thorough assessment and remedial actions described above, needed preventive measures are taken:

- Create easier instruction displays, visual controls and mistake proofs.
- Prepare tentative work standards and procedures, and then follow them.

12.3.3 Period 2: Establish usage conditions of tools and jigs

After learning about tools, jigs, measuring apparatus, kanbans, and other relevant materials to be dealt with, operators conduct an overall inspection to improve usage conditions of tools and jigs as well as work methods in the following sequence:

- Learn about materials to be dealt with by operators.
- Know structure and function of products.
- Know objectives, quality, and requirements of human work.
- Conduct overall inspection of tools, jigs, measuring apparatus, visual controls and mistake proofs in terms of functions, structure, objectives, breakdowns, defects, ease of use, and better alternatives.

The results of the above activities are compiled into tentative standards. In addition, all materials residing on the shopfloor are totally assessed once again with an aim to attain just in time supply of materials by eliminating waste of stock and human motion, as listed here:

- Location and manner of storage and quantity of tools, jigs, and measuring apparatus
- Shape, function, and dimension of pallets, wheelers, vehicles, and other material handling equipment
- Shape and dimension of workbenches, parts shelves, and containers
- Methods and styles of packing, and quantity of packages

When different types of activities, involving mutually affecting tasks one with another, are proceeding in the multiple departments according to their own different programs, as for instance, in the assembly and material handling departments, a system for providing a common and synchronous supply/receipt of parts/kanban must be established through adequate cooperation of all related departments. Further details in this regard are illustrated in the following case studies, described in this chapter.

12.3.4 Period 3: Standardization and autonomous supervision

After reviewing the tentative standards prepared in the previous periods, operators finalize work standards and procedures. In other words, they improve software and hardware until the easiest, most consistent, and least wasteful manual work procedures are achieved. Finally, operators arrive at a status of autonomous supervision.

12.3.5 Two seven-step programs on the same shopfloor

When a certain manual work shopfloor has small-scale equipment, two different programs, in connection with equipment and manual work, are prepared for autonomous maintenance activities and developed on the same single shopfloor. Of course, where equipment is not so prominent, it is much easier to apply a single program in response to manual work-oriented TPM activities. Small pieces of equipment are dealt with by way of a short remedial program.

12.4 Autonomous Maintenance in the Assembly Department

12.4.1 The aims of autonomous maintenance

In the automobile industry, most assembly work still relies mainly on manual work, due to a wide variation in models, specifications, and the involvement of a large quantity of parts to assemble. Moreover, the recent severe shortage of labor as well as rapid fluctuations in demand for cars are compensated for by bringing in temporary operators from outside the company. This incoming and outgoing of operators, along with the frequent reallocation of personnel, results in an increase of manual work losses.

In view of this situation, a company decided that autonomous maintenance activities would be directed toward the realization of an

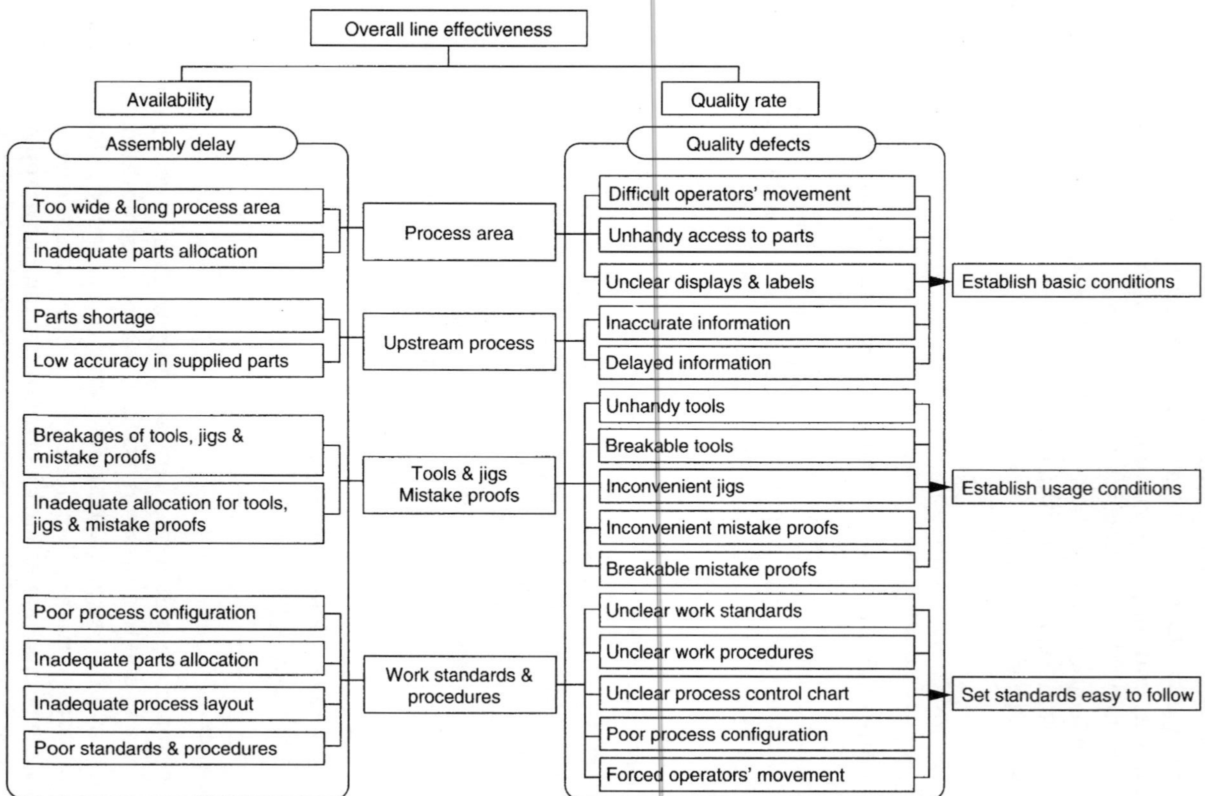

Figure 12.1 Overall line effectiveness and current problems.

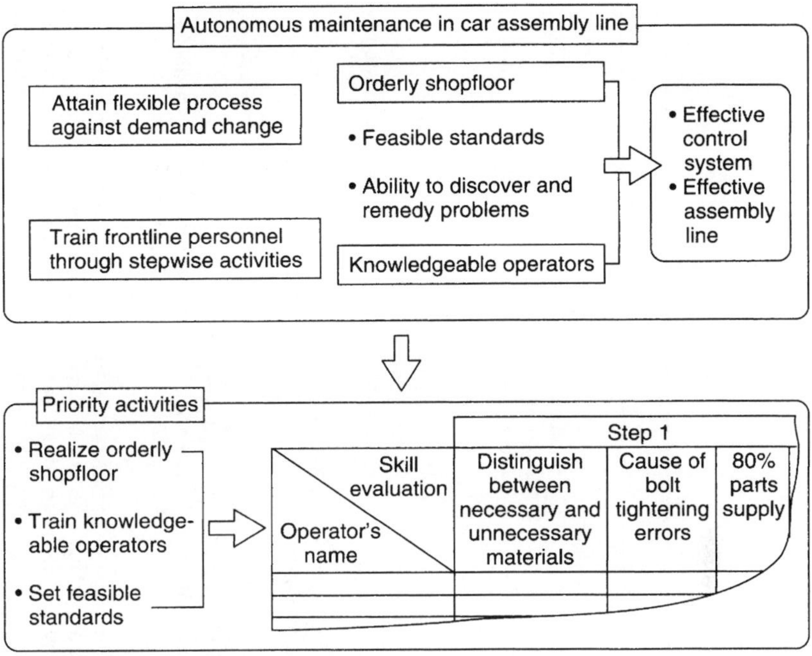

Figure 12.2 An approach toward autonomous maintenance.

orderly shopfloor in order to facilitate manual work for and decrease losses by newcomers. Figure 12.1 illustrates this approach in a car assembly plant. From a human perspective, training of knowledgeable operators who find and remedy any kind of abnormalities makes it possible for operators to undertake necessary action on their own initiative.

These knowledgeable operators in combination with an orderly shopfloor assure that productive, safe, and flexible processes are anticipated, as diagrammed in Fig. 12.2.

12.4.2 The seven-step program for assembly work

In response to a classification of recognized problems under present operating conditions, the seven-step program shown in Table 12.1 is constructed to repeat a sufficient number of CAPD cycles. Having established processes wherein abnormalities are detected at a glance, it is essential to proceed decisively with the following actions:

- Eliminate unnecessary materials and dropped parts from shopfloor.

TABLE 12.1 Autonomous Maintenance Program Developed in Car Assembly Line

	Step	Aims	Activities
Establish basic conditions	1. Initial cleaning	Learn "cleaning is inspection" through cleaning and removing unnecessary materials from work area.	▪ Thoroughly remove unnecessary materials. ▪ Conduct survey and analysis for dropped parts, and take remedial action. ▪ Clearly display storage area for parts, tools and jigs.
	2. Tidying-up process	Look for ease of reading and workability through tidying up work area.	▪ Display legible labels for parts shelves. ▪ Modify parts shelves for ease of use. ▪ Review installation methods for parts containers.
	3. Tentative standards	Set tentative standards to maintain process conditions attained in Steps 1 and 2.	▪ Set tentative cleaning and inspection standards. ▪ Thoroughly follow rules set by operators themselves.
Establish usage conditions and work methods	4-0. Product characteristics education	Learn about structure and function of product to attain quality assurance in car assembly work.	▪ Identify the worst five quality problems occurring in process allocated, and learn about quality by solving relevant problems.
	4-1. Overall tool inspection 4-2. Overall jig inspection 4-3. Overall measuring apparatus inspection 4-4. Overall mistake proof inspection	Conduct overall inspection in terms of hand tools, jigs, measuring apparatus and mistake proofs to assure quality results.	▪ Obtain optimal tools, jigs, measuring apparatus and mistake proofs to assure quality results. ▪ Obtain easy-to-use tools and jigs along with easy-to-work, safe processes so as to attain accuracy and durability in manual work.

TABLE 12.1 *(Continued)*

Standardize	5. JIT parts supply	Remove wastes from inventory and human motion.	▪ Review parts shelves to make them optimal. ▪ Review stocked parts to attain optimal inventory control. ▪ Review allocation of materials and human motion in and around assembly line to search for optimal combination.
	6. Standardization	Review and improve tentative standards to make them easier to follow.	▪ Compare in detail tentative standards with given work procedures to finalize operators' routine work standards.
	7. Standard work sequence	Implement standard work sequences to remove qualitative variability.	▪ Establish standard work systems. ▪ Attain flexible process against variablity of production output.

TABLE 12.2 Definition of Unnecessary Materials and Dropped Parts

Unnecessary materials	Dropped parts
▪ Unused parts, hand tools, jigs or measuring apparatus	▪ Parts dropped from workbench
▪ Unlabeled parts or any other materials	▪ Parts dropped from parts container
▪ Unlabeled temporarily stocked materials or pallets	▪ Parts dropped due to assembly errors
▪ Materials left directly on floor	▪ Wrapping paper, protective packing or base paper

- Assess the layout of processes along the assembly line.
- Take countermeasures to the deterioration of and strive for easier use of tools and jigs.
- Supply parts to assembly line on a just in time basis.
- Prepare work standards, procedures, and manuals.

For the sake of clarity, the definition of unnecessary materials and dropped parts is examined in Table 12.2.

12.4.3 Case study

Step 1: Initial cleaning. In a loom assembly plant which has four production lines, 164 operators are working on a single daylight shift basis. Initial cleaning is carried out in the following sequence:

1. Clean shopfloor of dirt and dust.
2. Pick up dropped parts and assess their occurrences.
3. Identify unnecessary materials. (Attach identification tags.)
4. Eliminate unnecessary materials. (Remove identification tags.)
5. Detect deteriorated and defective tools and jigs. (Attach identification tags.)
6. Remedy deteriorated and defective tools and jigs. (Remove identification tags.)

Due to the long history of the factory, there is a large quantity of unnecessary materials. An example of the thorough elimination of such dispensable materials is illustrated in Fig. 12.3.

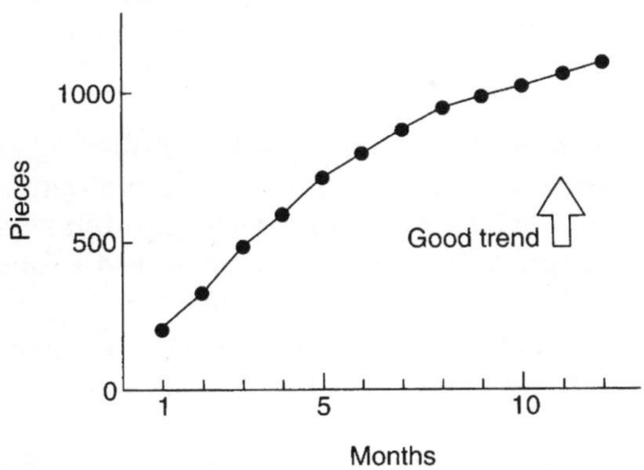

Figure 12.3 Unnecessary materials discovered.

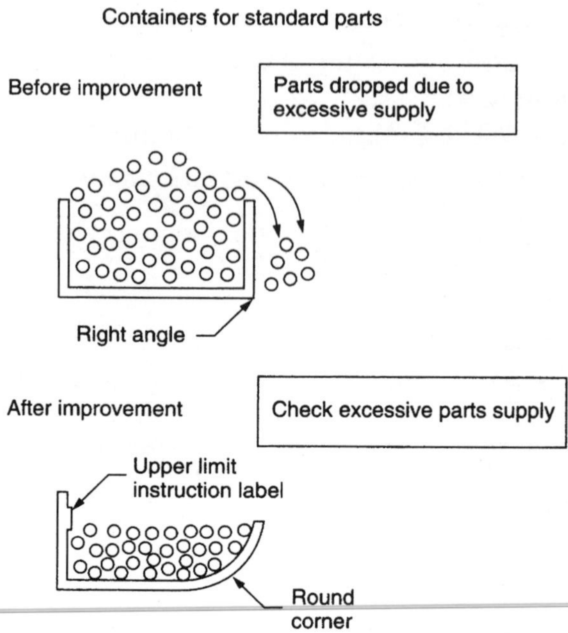

Figure 12.4 The remedial action taken for dropped parts.

Step 2: The tidying-up process. On a shopfloor in the same plant just described, required remedial actions are taken to eliminate the sources of contamination along with unnecessary materials and dropped parts, and to modify difficult cleaning areas. Successful results thereby obtained in the managers' models, which are applicable to other processes, are edited into one-point lessons for educational aids for operators. During the following span of about a half year, operators took 323 remedial actions against 193 difficult cleaning areas, 72 sources of contamination, and 58 other problems. Typical examples are shown in Figs. 12.4 and 12.5.

Step 3: Tentative standards. Partitioning and numbering systems for the floor and shelves are tentatively standardized and applied to entire production lines. Floors, which previously were painted randomly, are now painted with regard to standards for color codes and partition lines. Recognizing that it is important for operators to set and follow their own rules, they complete the establishment of basic operating conditions. Figure 12.6 is a sample of tentative standards.

Step 4: Overall inspection. To begin with, operators were taught the characteristics of the products manufactured by themselves. They then

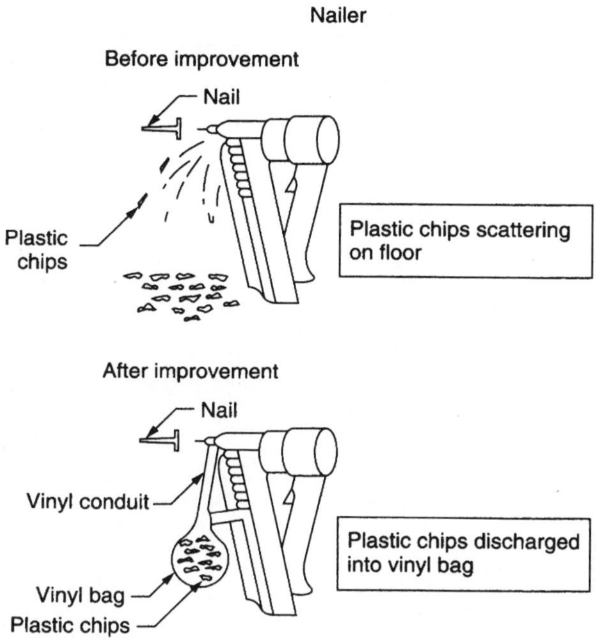

Figure 12.5 The remedial action taken for a source of contamination.

conducted an overall inspection in accordance with specified categories such as tools, jigs, measuring apparatus, and mistake proofs in terms of accuracy and continuity of manual work.

Overall tool inspection. Operators made a systematic assessment of work methods and tools with which they always had difficulties and experienced inconvenience. Figure 12.7 details major checkpoints and the results of the overall tool inspection. By means of this type of thorough inspection, many problems could be pinpointed. Figure 12.8 illustrates a typical modification of a hand tool.

Lube pipes used to be assembled onto gear boxes installed in a car by using mass-produced, open-end wrenches. Oil leakage sometimes followed due to insufficient tightening torque. Operators responded by inventing their own easy-to-tighten wrenches, which also simultaneously confirmed the proper torque. Such an idea never originated with an engineer because it required a kind of wisdom born on the shopfloor.

Overall jig inspection. Operators checked the problems in jigs in order to develop easy-to-use jigs and to facilitate their manual work. An

Tentative autonomous maintenance standards	Line	PM group	Managers' group	Standard No.	A-001
		All lines	Process Code No.	Date prepared	
			——	Date issued	

Numbering system

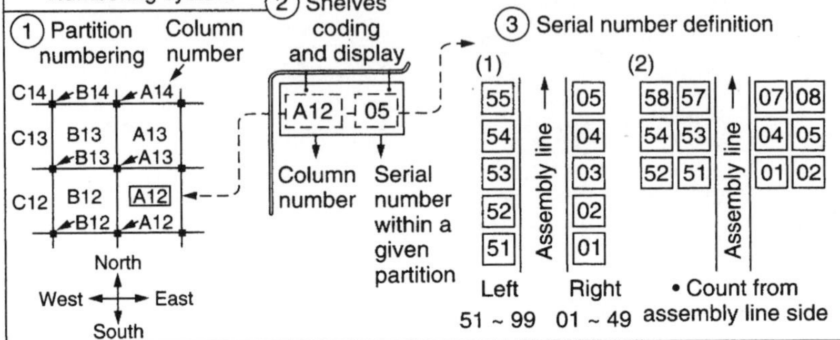

No.	Item	Standards	Relevant standards	Installed by	Checked by
1	Partition numbering	• North and east floor of each column is defined as partition • Partition number is labeled by column number		Supervisor	Manager
2	Shelves coding and display	• Shelves code number is made up of hyphenated three digits of column number and two digits of serial number Example: B12 - 05 (Column - Serial number)		Supervisor	Manager
		• Assign 01 ~ 49 to right side and 51 ~ 99 to left side of assembly line			

Figure 12.6 Tentative standards for a numbering system.

example of this activity is shown in Fig. 12.9. A steering gear (weight 7.0 kg) used to be supported by the left knee while it was assembled with both hands into a car. But fatigue in the foot and the time spent for adjusting the spline of the shaft sometimes resulted in assembly line stoppage. In response to this problem, a jig synchronized with assembly line speed was invented in order to vertically adjust the gear position by pedaling with the left foot. An accurate and easy assemblage was thereby achieved. Assembly line stoppage was reduced to Zero.

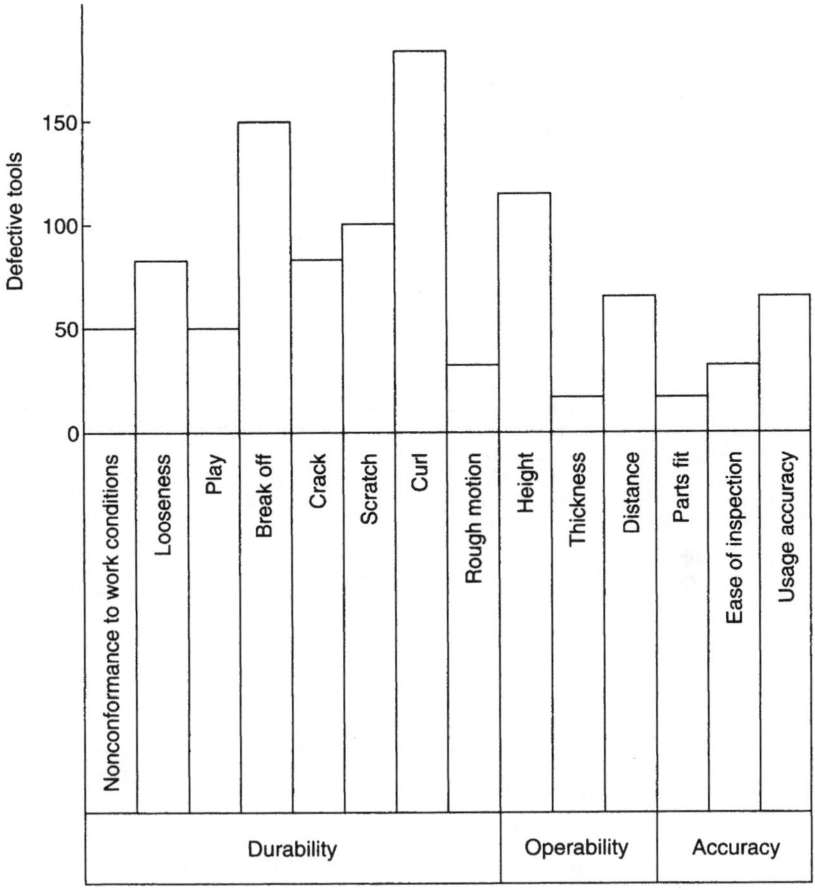

Figure 12.7 Overall tool inspection.

Step 5: Just in time parts supply using kanban. To achieve just in time parts supply to assembly lines, methods for and timing in acceptance of parts were thoroughly examined. As the results of many remedial actions, both the quantity of parts accepted and the standard stock of parts were minimized. During this activity, problems shown in Fig. 12.10 were detected and remedied. An example of remedial actions taken for acceptance of parts is illustrated in Fig. 12.11. As a result of the effective utilization of kanban, standard stock of parts were reduced 40 percent.

Reversed idea: movable workbench. By observing a loom subassembly line, waste was detected in the operators' movement, pickup of parts, and press work. According to a more detailed analysis of work meth-

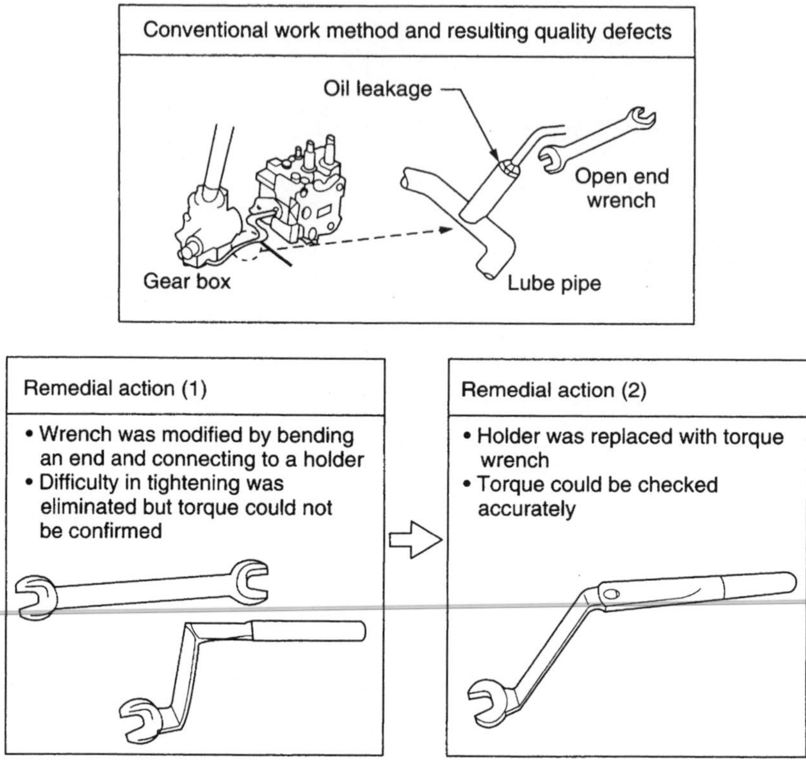

Figure 12.8 How operators improve their own tools.

ods, the problems were found to be associated with the location of the workbenches, parts, and hand tools.

The process was then modified, as demonstrated in Fig. 12.12. Workbenches, traditionally fixed at each suitable location on the work floor, were modified to be able to travel on guide rails in accordance with the progress of assembly work and operators' motion. An extensive application of this reversed idea to other similar processes resulted in a reduction of 30 percent of assembly time and 18 m^2 of floor space.

Step 6: Standardization. In reference to a compressor assembly line for automobile air conditioners, a series of study meetings took place to make operators' work easier. Consequently, the assembly sequence, tools, jigs, and parts pickup were improved based on a detailed analysis of the roles and motion of operators, especially their body, hands, feet, and eyes, during the assembly work. Figure 12.13 details the problems detected in the operators' motions.

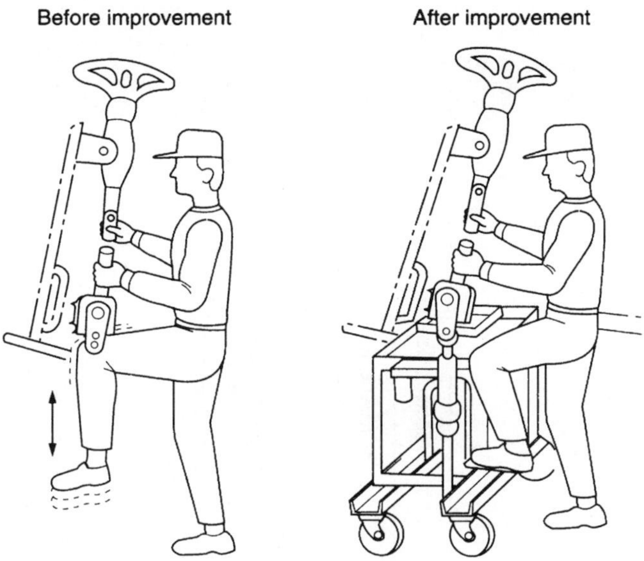

Figure 12.9 Facilitating steering gear assembly work.

The nature of a delay in a parts assembly line was not clearly ascertained because the spaces between workpieces on conveyors were obscure. It resulted in 1500 minutes of line stoppage per month in entire assembly lines. To solve this problem, workpieces were made to stop at fixed points with regular spacing as an application of visual controls. This innovation, in conjunction with a modification of the lamps that display current operating status, cut the assembly line stoppage by half, as illustrated in Fig. 12.14.

12.5 Autonomous Maintenance in the Inspection Department

12.5.1 The aims of autonomous maintenance

In many companies, inspection work presents various challenges. For example, in the automobile industry:

- Quality characteristics to be inspected for are diverse in regard not only to essential mechanical functions, but also to minute exterior finish.
- Demand for quality becomes more rigorous yearly.
- The same quality defects are repeated several times.

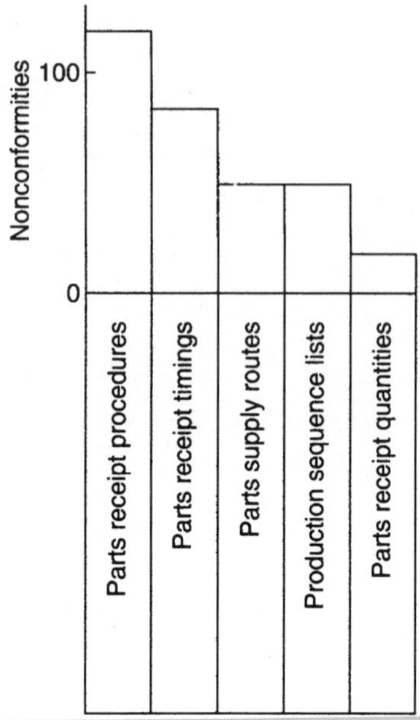

Figure 12.10 The nonconformities discovered in activities to attain JIT parts supply.

- Some quality defects are overlooked by inspectors' who lack proficiency because they are repeatedly reallocated according to frequent changes in production plans.

Further investigation often reveals the following causes of problems:

- Poor circumstances surrounding the inspection processes, for example, lighting, layout and configuration of a process, tools, measuring apparatus, inspection standards, and criteria
- Insufficient inspectors' knowledge about detailed structure and the function and quality characteristics of a product
- Inadequate inspection standards, manuals, check sheets, and quality samples

In view of these factors, inspection work for engines was assessed and the six big inspection losses were highlighted, i.e., inspection errors, evaluation errors, breakdowns in measuring apparatus, waiting time for necessary tests and their evaluation results, and insufficient inspectors' skill. On the basis of this preliminary survey, major targets and goals of autonomous maintenance activities were set for the elimination of these six losses in order to achieve an optimal

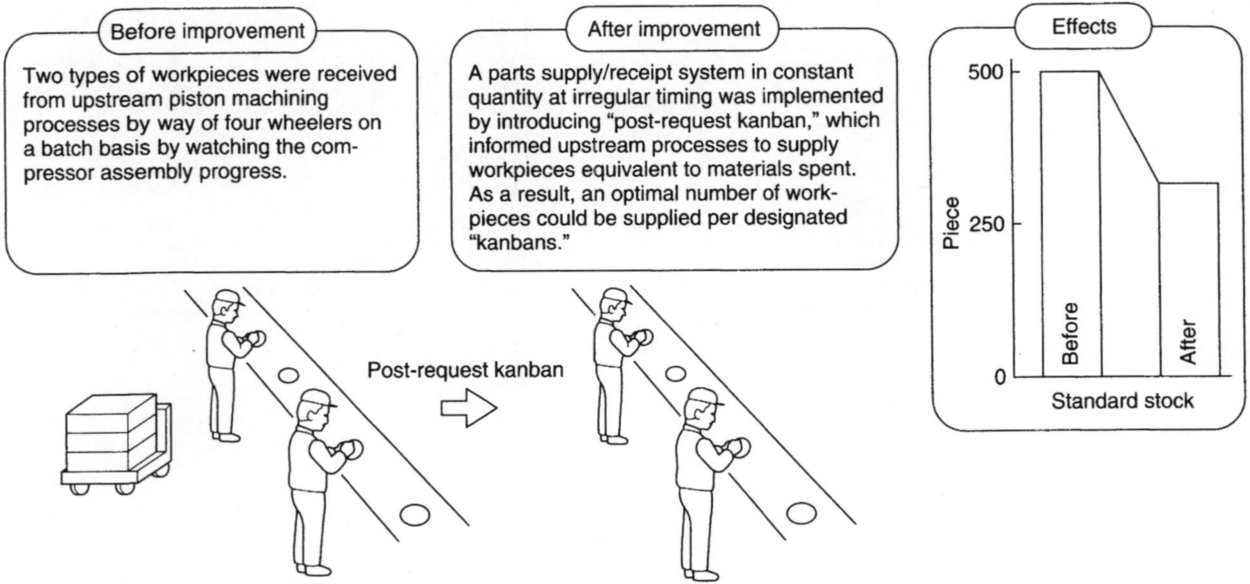

Figure 12.11 Improving parts' receipt methods.

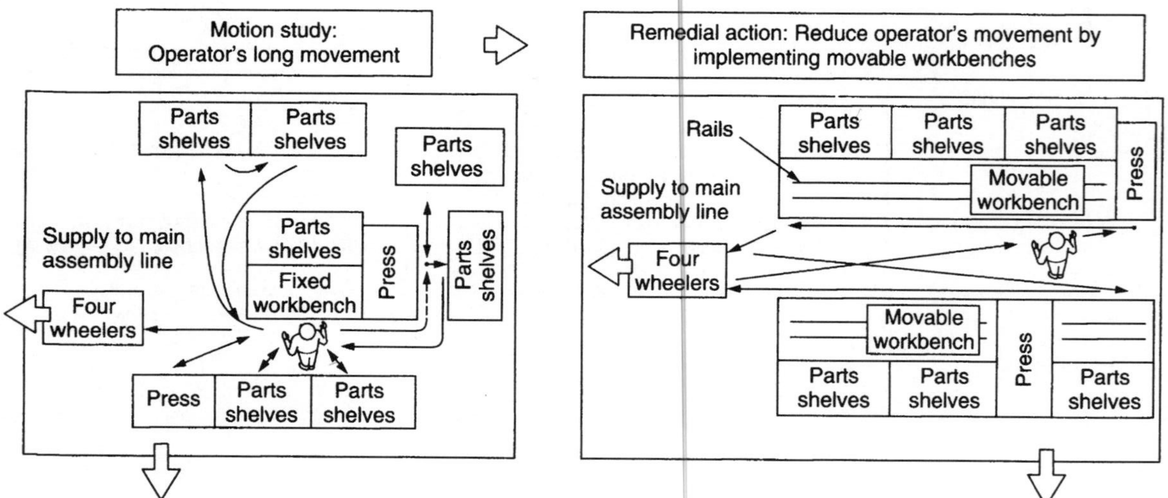
Motion study:
Operator's long movement
Parts shelves
Parts shelves
Parts shelves
Parts shelves
Fixed workbench
Press
Parts shelves
Supply to main assembly line
Four wheelers
Press
Parts shelves
Parts shelves
Remedial action: Reduce operator's movement by implementing movable workbenches
Rails
Parts shelves
Parts shelves
Parts shelves
Movable workbench
Press
Supply to main assembly line
Four wheelers
Movable workbench
Press
Parts shelves
Parts shelves
Parts shelves

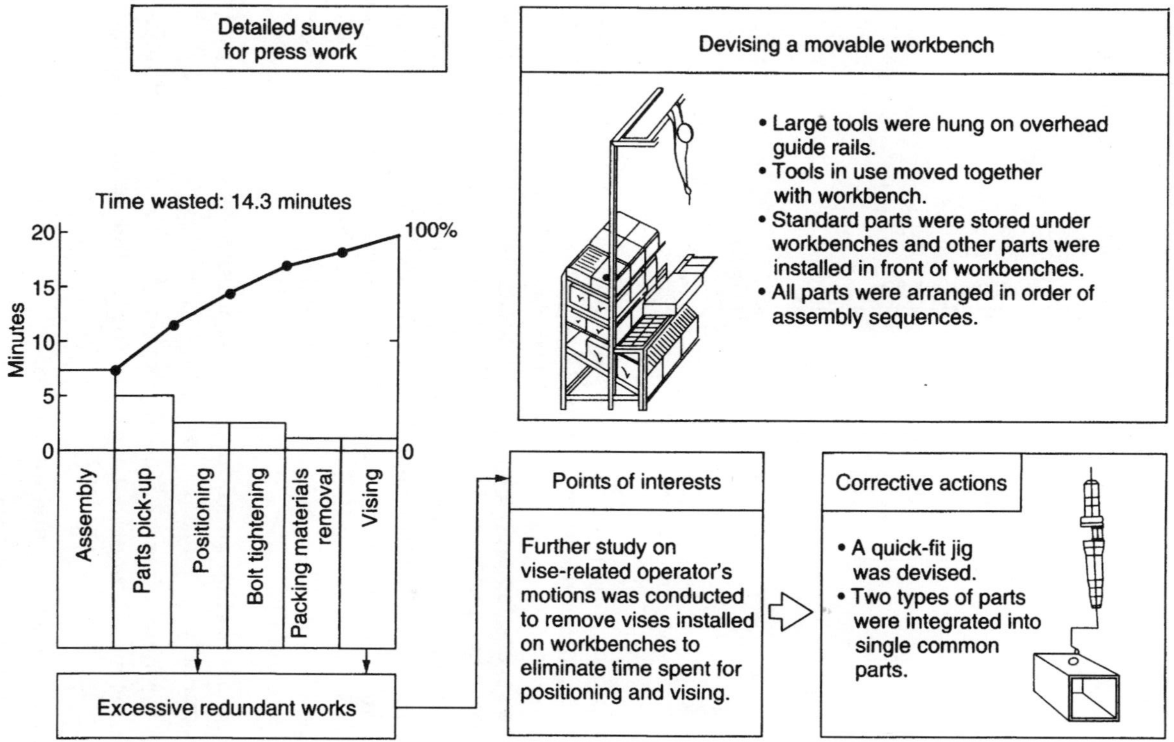

Figure 12.12 Improving loom assembly processes

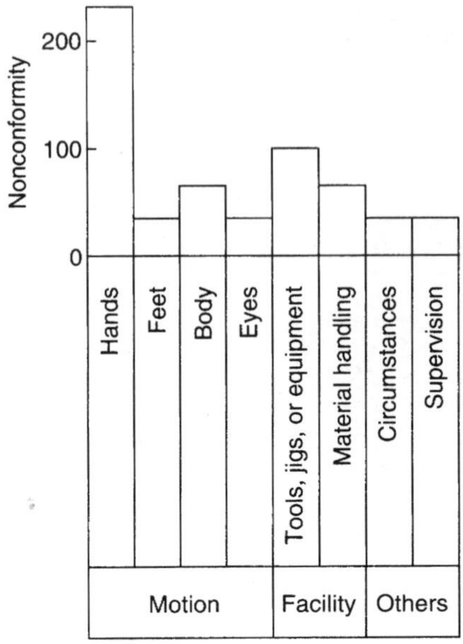

Figure 12.13 The nonconformities discovered in operators' workability.

inspection process and to train inspectors knowledgeable about quality as well as the mechanisms of products.

12.5.2 The seven-step program of autonomous maintenance

The seven-step program shown in Table 12.3, in response to the results of the survey, called for the repetition of the CAPD cycle. By developing this kind of program, an easier and more reliable inspection process, with related inspection standards and the training of more knowledgeable inspectors at quality, is expected.

Some more details:

- Establish inspection procedures in accordance with an inspection work sequence.
- Identify problems in terms of inspection by way of the established procedures and categorize them into the six big losses, as shown in Table 12.4.
- Remedy the problems just mentioned.
- Prepare tentative inspection standards.
- Review three major elements of inspection (inspection standards, measuring apparatus, evaluation criteria) by learning the structure and function of products.
- Standardize these three major elements of inspection.

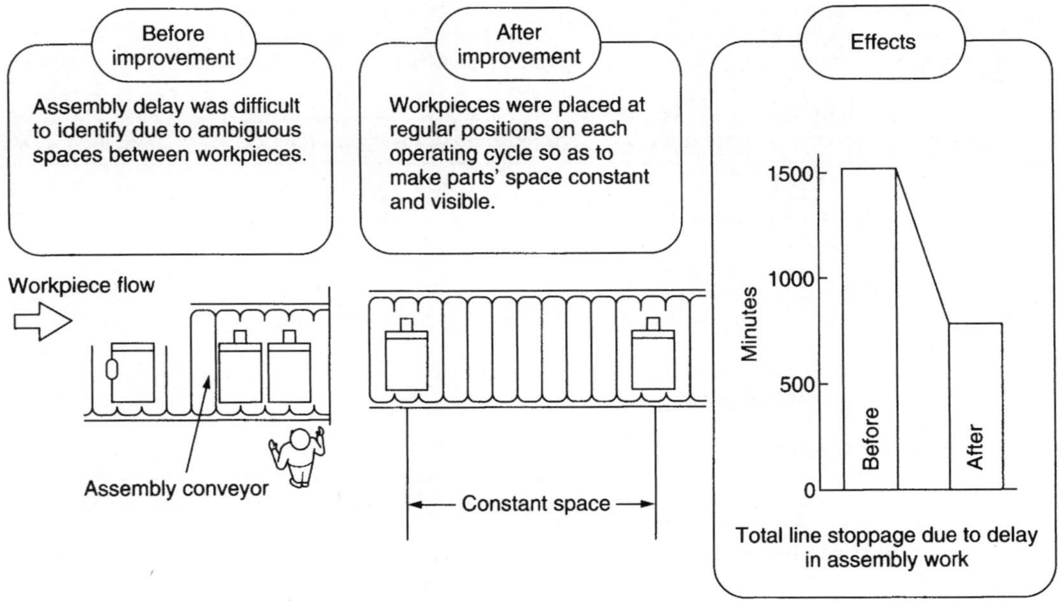

Figure 12.14 Stopping workpieces at regular positions on the conveyor (visual control).

TABLE 12.3 Autonomous Maintenance Program in Inspection Department

	Step	Aims	Major activities	Six big inspection losses
Establish basic conditions	1. Inspecting work assessment	Assess inspection procedures, work methods, instruments and inspection data.	▪ Locate defective areas and pose questions. ▪ Remove unnecessary inspection data. ▪ Discover defects in inspection equipment and instruments.	Inspecting errors
	2. Inspecting work improvement	Restore and correct defective areas.	▪ Restore and remedy deteriorated and defective parts in inspection equipment. ▪ Establish clear and easy filing methods for inspection data.	Evaluating errors Inspecting equipment breakdowns
Observe usage conditions	3. Inspecting work standards	Set tentative inspection standards and procedures.	▪ Set tentative standards to maintain conditions attained in Steps 1 and 2.	
	4. Overall inspection	Conduct inspection education for product characteristics and subsequent overall inspection for automotive inspection work.	▪ Learn about structure, function and quality characteristics of products. ▪ Discover and remedy defects in instruments. ▪ Check conformity of accuracy of instruments with quality specifications. ▪ Facilitate setup work for inspection. ▪ Review inspection check sheets.	Evaluating time

TABLE 12.3 *(Continued)*

	5. Quick data handling	Reduce inspection time.	To realize accurate and nonwasteful inspection: ▪ Review inspection standards, equipment and evaluation criteria. ▪ Review inspection methods and quality assurance criteria.	Reliability testing time
Standardize	6. Standardization	Definitely follow rules.	▪ Finalize clear inspection standards that are easy to understand and follow.	Inspection skills
	7. Autonomous supervision	Improve further.	▪ Accumulate improvement by way of thorough implementation of visual controls.	

TABLE 12.4 Evaluating Inspection Work in Terms of Six Big Inspection Losses

		Major inspection work	Frequency	Work summary	Six big inspection losses					
					Inspection errors	Evaluation errors	Inspecting equipment breakdowns	Long evaluation time	Long reliability testing time	Insufficient inspection skills
Parts inspection	1	Detailed sampling inspection	2/D	Make detailed precision inspection of areas (surface finish, shape, out-of-roundness, and straightness) unable to be dealt with by operators and advise production department if needed.	○	○		○		○
Performance test	1	Sampling performance test	1/W	Evaluate performance and finish of tuned-up engines, and plot data into quality control charts.	○	○		○	○	
	2	Inspection equipment check	1/M	Conduct routine and periodic check of inspection equipment.			○			

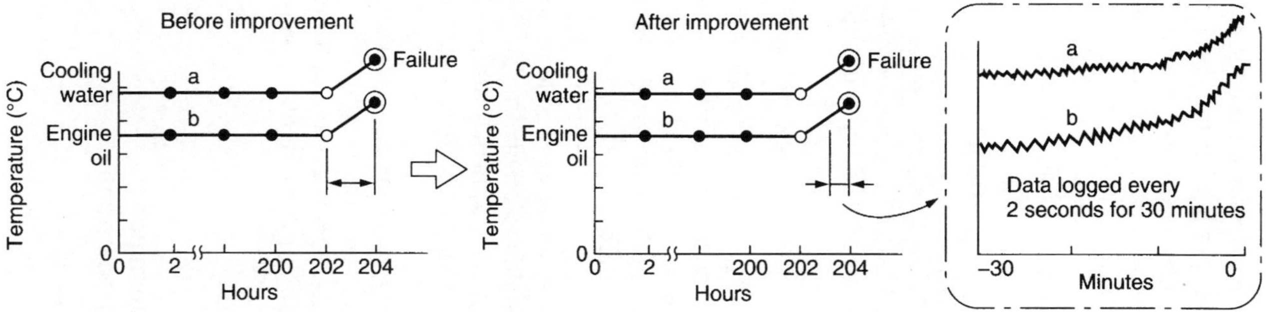

Figure 12.15 Improving a data logging system.

12.5.3 Case study

Step 2: Assess inspection work. When an endurance test of an engine is carried out under very severe load simulating various users' driving conditions, failures sometimes occur in the engine. To determine the cause of failure, relevant data such as hydraulic pressure, torque, combustion temperature, and cooling water temperature are needed. The conventional endurance test, however, did not record these data prior to the occurrence of a failure, so it became necessary to repeat the same test from the beginning.

In order to solve this problem, a computer based data logging system was developed by which 14 parameters were kept every 2 seconds for 30 minutes, as illustrated in Fig. 12.15. The cause of a failure, therefore, was always identified at an unexpected cessation of a test. As a result, some hundreds of hours for additional endurance test runs were unnecessary.

Step 4: Overall inspection. The annual increase in types and models of engines led to a longer time for mounting the engine onto the test bench with proper alignment between the engine to be tested and the dynamometer. Inspectors undertook to eliminate these setup and adjustment losses in the endurance test by way of replacement of brackets and changes in support positions and alignment. Consequently, some 30 related problems were also remedied.

One of the typical examples is illustrated in Fig. 12.16. The engine was formerly mounted onto a test bench after being brought into the laboratory. Inspectors, instead, developed a transportable test bench provided with supports to accommodate an engine prior to arrival at the laboratory. Accordingly, another engine was set up in a preparation room during a test run. This resulted in a reduction of 70 minutes for the setup of an engine. Over a month, 1500 minutes spent for setups were reduced to 525 minutes. In other words, a 63 percent time reduction was achieved.

12.6 Autonomous Maintenance in the Material Handling Department

12.6.1 The aims of autonomous maintenance

The material handling department is in charge of supplying raw material and component parts to assembly lines, and providing transportation of materials between processes. Because recently growing car

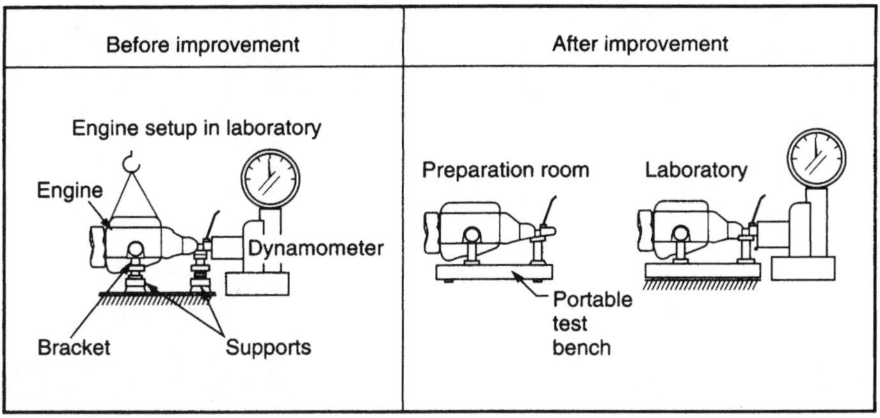

Figure 12.16 Improving the setup methods of engines.

demand and a variety of models caused a rapid increase in the types of materials handled, the following problems arose:

- Increasing a number of very similar parts results in mistakes and delay in parts supply to assembly lines.
- Diverse parts specifications result in the expansion of the storage area.

In response to this situation, major targets of autonomous maintenance activities are set for the creation of a parts supply system flexible enough to accommodate diverse parts and rapid fluctuations in production output.

12.6.2 The seven-step autonomous maintenance program

The seven-step program, as listed in Table 12.5, is structured to sufficiently repeat the CAPD cycle. By developing this program, the thorough elimination of waste in parts stock, storage areas, and human motions is expected.

12.6.3 Case study

Step 2: The tidying-up process. In order to prevent parts supply errors, baffle plate-style "swinging kanbans" were attached at the walkway side of parts shelves installed along car assembly lines, as illustrated in Fig. 12.17. This arrangement assures that anyone immediately understands what parts are to be supplied at which point on shelves.

TABLE 12.5 Autonomous Maintenance Program in Material Handling Department

	Step	Aims	Activities
Establish basic conditions	1 Initial cleaning	Learn "Cleaning is inspection" by removing unnecessary materials from work areas.	▪ Thoroughly remove unnecessary materials. ▪ Take corrective actions to sources of dirt and dust. ▪ Do not place parts directly on floor without proper protection or outside designated storage areas.
	2 Work area tidy-up	Seek legibility and workability through tidying up work areas.	▪ Assess empty container return and parts storage methods. ▪ Assess parts supply methods into storage in assembly lines. ▪ Standardize displays and labels.
	3 Tentative standards	Set tentative standards to maintain work area conditions attained in Steps 1 and 2.	▪ Establish adequate systems and easier-to-follow rules with certainty. ▪ Set standards to supply specified amounts of parts to assembly lines. ▪ Eliminate temporary storage and carry-back.

TABLE 12.5 *(Continued)*

	Step		Objective	Activities
Establish usage conditions and work methods	4	4-0. Kanban education 4-1. Parts education 4-2. Overall strap inspection 4-3. Overall pallet inspection 4-4. Overall parts supply inspection	After learning about kanban and materials to handle, conduct overall inspection in terms of kanban, strap, pallet and parts supply method.	▪ Learn about kanban such as circulation cycles, quantitative specification, purchasing routes and delivery days. ▪ Learn about parts to handle such as structures, function, initial parts, priority car parts and mis-supplied parts. ▪ Understand parts distribution system. ▪ Understand parts receipt and shipping work. ▪ Pursue optimal parts supply to lines as follows: Can lots be made smaller? Are supply methods easy? Are there no defects?
	5	JIT parts supply	Remove waste from materials in stock and manual work.	▪ Set adequate quantity of parts supply and stock adapted to production quantity by reviewing: Allocation of parts shelves. Number of pieces of kanban and timing to ship. Material handling tools, wheelers and pallets. Style and quantity for parts transportation.
Standardize	6	Material handling standards	Review tentative standards to make them easier to follow.	▪ Set standards to attain easier parts reciept and supply. ▪ Establish parts supply methods fit production cycle time.
	7	Autonomous supervision	Implement standard work procedures and reduce quantitative variability of materials in stock.	▪ Establish standard work procedures. ▪ Attain flexible material handling systems against changes in production output.

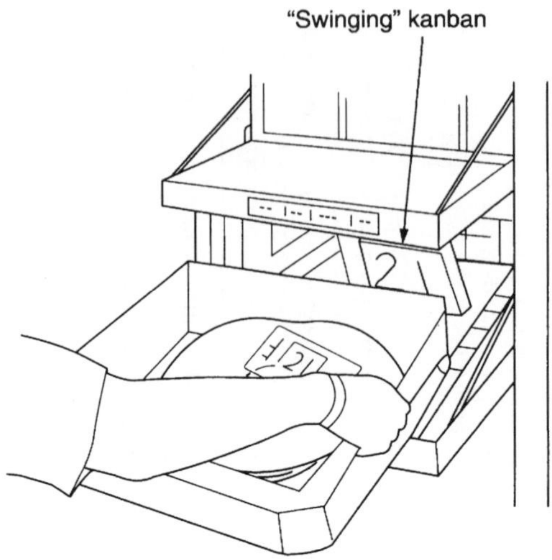

Figure 12.17 Preventing parts' supply mistakes.

Step 4: Overall inspection. After learning about kanban and the materials transported, operators conducted an overall inspection of pallets, kanbans and their straps, and parts supply methods. As a result of these thorough inspections, many problems were exposed and suitable remedial actions were taken. Two examples are examined.

A large number of posts, which were installed along the car assembly line to temporally store kanbans to be collected by material handling personnel, presented various difficulties not only to assembly work, but also to the collection of kanbans. To solve this problem, a chute was attached at the sides of parts shelves connected to kanban posts, which were relocated on the walkway side of the assembly line, as illustrated in Fig. 12.18. These remedial actions reduced the time

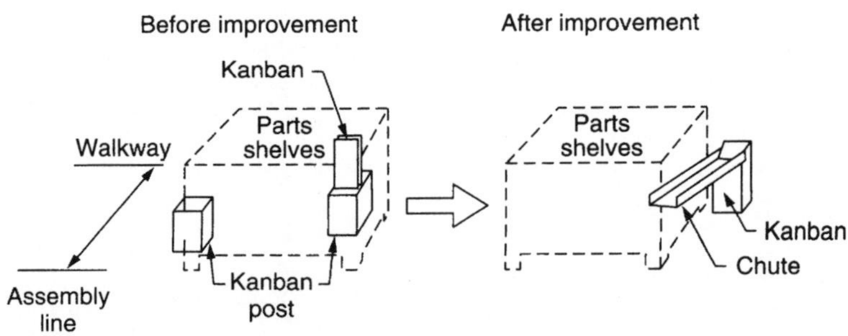

Figure 12.18 Reducing kanban recovery time.

required for collecting kanbans by half and, as a side effect, collisions between material handling personnel and car assembling operators also were avoided.

The second example paints the long time required to stack the finished engines onto pallets because no guide for doing so was ever provided. As a solution, both the edges of pallets were painted with a white color. Such a minor idea resulted in a remarkable reduction of stacking time.

Glossary

activity board An information board prepared by each PM group to facilitate operators' communication and mutual understanding, and display the performance of group activity to other frontline personnel.

automation The application of automated equipment in eliminating and/or facilitating various types of manual work by means of mechanical, electrical, pneumatic, hydraulic, or computer-aided control systems.

autonomous maintenance Operators' step-by-step activities when searching for the optimal conditions of operations and production (an orderly shopfloor) by way of the establishment of basic equipment conditions (cleaning, lubrication, and tightening), and, as a result, the arrival at an autonomous supervision (operators' self-control) environment.

autonomous maintenance audit An audit for PM group activities conducted by frontline managers and engineers at the end of a given step wherein equipment conditions along with each operator's performance are evaluated.

autonomous maintenance standards Operators' routine maintenance standards set in Step 5 comprised of a combination of cleaning/lubricating standards set in Step 3, and categorized inspecting standards set in Step 4.

bench mark The initial conditions of major indexes which are assessed prior to the commencement of TPM activities in connection with the six big losses and PQCDSM, and which are applied to evaluate the progress in operations and production.

CAPD cycle A series of remedial actions in TPM, check (C), act (A), plan (P), and do (D), to be applied when any given autonomous maintenance program is planned.

chronic loss The gap between actual equipment effectiveness and its optimal value when the same loss recurs in a narrow range of incidence.

cleaning is inspection Cleaning in TPM is not conventional, industrial housekeeping, but entails thorough cleaning in order to expose all hidden defects in equipment and restore it properly.

cleaning/lubricating standards Operators' routine cleaning and lubricating standards made up of a combination of cleaning standards set in Step 2 and lubricating standards set in Step 3, whenever the given time targets are attained in Step 3.

cleaning standards Operators' routine cleaning standards tentatively set at the beginning of Step 2 to maintain the cleanliness achieved in Step 1, and revised at the end of Step 2 with reference to the remedial actions taken.

color lubrication control Visual lubrication and lubricant control in order to facilitate routine lubricating tasks, and avoid human errors.

complex causes Plural causes in a piece of equipment which individually do not affect productivity and product quality, but do result in losses when a particular combination of these same causes comes into existence simultaneously and by chance.

computerization The application of computers in eliminating and/or facilitating various types of manual work as well as conventional control systems of equipment.

defective (off-specification) product A product or workpiece which does not meet given quality control standards (quality specifications).

easy-to-manufacture product design The incorporation of easiness in manufacturing from planning through detailed design during the engineering stage of a forthcoming product.

elimination of the six big losses Activities developed generally on the basis of project teams organized by personnel drawn from the production, maintenance, plant engineering, and any other relevant departments and searching for the elimination of the six big losses.

exposed defects Defects in a piece of equipment or its parts which may be easily identified at a glance.

exposed loss An operating condition which can be recognized at a glance as loss by anyone.

five criteria for ease of observation Five criteria to evaluate human motion and performance in observing a given rule; i.e., (1) Is a rule clearly defined?, (2) Is a rule well understood?, (3) Is a rule well observed?, (4) Is a rule easy to observe?, and (5) Are deviations detected at a glance if a rule is violated?

five criteria for quality assurance Five criteria to evaluate quality conditions built into a piece of equipment in order to assure process quality; i.e., (1) a quality condition is quantitative or clear, (2) a quality condition is easy to set, (3) a quality condition resists variation, (4) a change in a quality condition is easy to detect, and (5) a change in a quality condition is easy to restore.

5W's and 1H Essential factors which must be incorporated into any given rules to be followed on manufacturing shopfloor.

forced deterioration The deterioration of a piece of equipment or its part progressing more rapidly than its inherent deterioration characteristics due to certain human performances.

four lists Generic term applied to the four lists onto which fundamental information is recorded throughout the development of the TPM seven-step program to prevent the neglect of any necessary matters; i.e., (1) defective area list, (2) question list, (3) source of contamination list, and (4) difficult work area list.

full-time maintenance The maintenance activities carried out by the maintenance department in contradistinction with operators' autonomous maintenance.

function-loss breakdowns Operating conditions under which breakdowns or malfunctions stop or impede the function of equipment and result in the cessation of production.

function-reduction breakdowns Operating conditions under which low-speed operations or quality defects occur, although production can continue.

hidden defects Defects in a piece of equipment or its part which are invisible for certain physical reasons or remain undetected due to a lack of proper human vision or attitude.

hidden loss A loss that goes unrecognized under seemingly normal operating conditions, and during the apparent maintenance of equipment effectiveness and product quality.

identification tag A tag tightly hung on or attached to problematic spots of equipment in order to clearly indicate and record the minimal information needed in terms of an identified problem and location, the corrective actions taken, and the results of such actions.

just in time (JIT) A state which assures that the necessary quantities of necessary materials exist whenever needed.

kanban (signboard) A tag or plate which bears the minimal information needed in terms of workpieces conveyed between processes and in terms of purchased raw materials.

knowledgeable operator An operator who is informed about matters of equipment as well as product quality.

lubricating point Point at which to supply lubricant.

lubricating standards Operators' work standards for routine lubricating task are tentatively set in Step 3 after finishing overall lubrication inspection in order to maintain the proper lubrication of equipment, and are revised in reference to the remedial actions taken.

lubricating surface Sliding parts which require lubrication such as reciprocating and rotating parts, and chains.

lubrication control system Routine lubrication conducted by operators or maintenance personnel in order to maintain proper lubrication of equipment by means of supplying lubricant, which is clean and conforms to operating conditions, to designated lubricating points and surfaces in the proper quantity and at the prescribed times.

maintenance prevention (MP) information Any useful information identified on the shopfloor in order to make maintenance unnecessary.

major defect A single defect in a piece of equipment that can cause its breakdowns and operation stoppages.

managers' model A suitable number of pieces of equipment to be dealt with by small groups, which consist of frontline managers (the managers' group), in order to experience and demonstrate an instructive example by practice with the equipment installed on their own shopfloor.

manual work losses Any kinds of losses, in terms of human errors and product quality, which exist on manual work shopfloors.

medium defect A single defect in a piece of equipment that can reduce its function but allows for continuing operation.

minor defect A single defect in a piece of equipment which cannot cause losses by itself, but results in losses only when a particular combination of these single defects occurs by chance.

misoperation information Any useful information acquired on the shopfloor in order to prevent the recurrence of an operator's misoperation.

multiple causes Plural conditions in a piece of equipment existing simultaneously, but producing effects independently, which can result in losses in terms of productivity and product quality.

natural deterioration The deterioration of a piece of equipment or part progressing in accordance with its inherent deterioration characteristics in proportion to the passage of time.

one-point lesson A sheet of paper on which only one subject is clearly described; also called the one-subject/one-sheet training method.

orderly shopfloor A shopfloor where any aberration from normal conditions can be detected at a glance by anyone.

overall inspection Inspection of both comprehensive and definitive steps applied to selected categories which are widely applied throughout a plant, such as lubrication, fastener, power transmission, electrical, instruments, hydraulics, pneumatics, water, steam, and so on. Lubrication is thoroughly inspected in Step 3. Other overall inspection categories are pursued in Step 4.

overlapped small group organization Small groups consist of five to seven people which are organized in accordance with corporate organizational structure. Lower-level leaders constitute an intermediate group, and have the role of connecting the overlapped, adjacent levels like a linking pin.

planned maintenance Activities which are carried out by the maintenance department in order to maintain and secure proper operating conditions of equipment by way of careful planning, prior scheduling, and preparation after establishing basic equipment conditions (cleaning, lubrication, and tightening), restoring defective parts, and maximizing parts' life.

PQCDSM The essential aspects to establish and evaluate TPM targets and performances; i.e., productivity (P), quality (Q), cost (C), delivery or inventory (D), safety (S), and morale (M).

preventive engineering Concepts and methodologies which eliminate, as early as possible, any potential seeds of future trouble anticipated to occur throughout the engineering, procurement, fabrication, installation, commissioning, and commercial production stages in connection with the manufacturing of a forthcoming product in a newly installed or revamped production line.

process quality assurance Clearly prescribed and effective maintenance of quality conditions incorporated into a piece of equipment or process in light of the five criteria for quality assurance along with ease of observation.

process quality Single or multiple elements of quality created and assured at each piece of equipment or process in response to product design.

quality (on-specification) product A product or workpiece which meets given quality control standards (quality specifications).

quality causes A certain number of factors maintained in an existing piece of equipment, creating the quality of a given product.

quality conditions A combination of a certain number of conditions which are incorporated into a piece of equipment in order to create the quality specified by the product design.

quality results The quality of a manufactured product which is created by an existing piece of equipment.

quality specifications Detailed elements of quality specified by product design.

roll-out education PM group leaders, educated beforehand by TPM instructors, teach acquired knowledge and skills to fellow operators involved in their own PM group.

second industrial revolution A trend of the times appearing in the late twentieth century wherein various types of manual work were substituted for equipment in response to the progress in computer and microelectronic technology.

short remedial program Operators' remedial actions programmed to resolve given problems from the six big losses recurring on their shopfloor, which are conducted, aside from the seven-step program, in a relatively short period of time.

single cause A single condition in a piece of equipment which can result in losses in terms of productivity and product quality.

six big losses Six major losses in reference to the inadequate condition of equipment operations and production; i.e., (1) breakdown loss, (2) setup and adjustment loss, (3) minor stoppage loss, (4) speed loss, (5) quality defect and rework loss, and (6) yield loss.

sporadic loss The gap between actual equipment effectiveness and its optimal value when the recurrence of a specific loss increases suddenly beyond the usual range of incidence.

TPM master plan Basic action plans and schedules which are prepared at company, division, department, and section levels in order to facilitate detailed planning of TPM activities at various layers and work places, and to encourage employees.

TPM office The administration office to coordinate various types of TPM activities which is installed at adequate corporate levels such as the entire company, a division, a factory, and a department.

visual control Surveillance control that uses a visual device which can reveal at a glance the normal or abnormal status of equipment operation or human behavior.

where-where analysis Careful observation to locate the precise spots in a given problematic area where the sources of problems exist.

why-why analysis Detailed observation to identify accurately the causes of problems, and give careful consideration to the remedial actions needed to prevent a recurrence of the same problems.

work order completion rate The ratio of the number of work orders submitted from the production department to the number of work orders completed by the maintenance department.

Zero-oriented concept A definite type of approach to searching for the absolute value, Zero, in human performances and equipment conditions, not by way of abstract armchair ideas, but by way of concrete and actual technical approaches.

Bibliography

Crosby, Philip B. *Quality Is Free.* New York: McGraw-Hill, 1986.

Gotoh, Fumio. *Equipment Planning for TPM: Maintenance Prevention Design.* Cambridge: Productivity Press, 1991. Reprint. Tokyo, Japan: Japan Institute of Plant Maintenance, 1988.

The Japan Institute of Plant Maintenance. *TPM Development Program: Implementing Total Productive Maintenance.* Cambridge: Productivity Press, 1989. Reprint. Tokyo, Japan: The Japan Institute of Plant Maintenance, 1982.

Juran, Joseph M., editor in chief, and Gryna, Frank M., associate editor. *Juran's Quality Handbook.* 4th ed. New York: McGraw-Hill, 1988.

Likert, Rensis. *New Patterns of Management.* New York: McGraw-Hill, 1961.

Index

ABOUT THE AUTHORS

MASAJI TAJIRI is an international engineering consultant, based in Tokyo and New York. He has worked as a systems analyst for engineering and construction services, and as a project engineer on large-scale oil refinery, gas processing, and chemical plant construction projects for Shell, BP, Exxon, and other companies.

FUMIO GOTOH is also an international engineering consultant, based in Tokyo. He has been a TPM consultant since 1973, and is a promoter of the "seven steps concept" in autonomous maintenance and the "four phases approach" to Zero Breakdowns. He has served as a consultant to Tokai Rubber, Topy, Toyota, Hitachi, Epson, and many other companies.